SpringerBriefs in Earth System Sciences

Series editors

Gerrit Lohmann, Bremen, Germany
Lawrence A. Mysak, Montreal, Canada
Justus Notholt, Bremen, Germany
Jorge Rabassa, Ushuaia, Argentina
Vikram Unnithan, Bremen, Germany

More information about this series at http://www.springer.com/series/10032

Michael Schulz · Andre Paul
Editors

Integrated Analysis of Interglacial Climate Dynamics (INTERDYNAMIC)

 Springer

Editors
Michael Schulz
Andre Paul
MARUM—Center for Marine
  Environmental Sciences and Faculty
  of Geosciences
University of Bremen
Bremen
Germany

ISSN 2191-589X   ISSN 2191-5903   (electronic)
ISBN 978-3-319-00692-5   ISBN 978-3-319-00693-2   (eBook)
DOI 10.1007/978-3-319-00693-2

Library of Congress Control Number: 2014953591

Springer Cham Heidelberg New York Dordrecht London

Printed on acid-free paper

Springer is part of Springer Science+Business Media (www.springer.com)

Contents

DFG Priority Research Program "Integrated Analysis of Interglacial Climate Dynamics" (INTERDYNAMIC)

Michael Schulz and André Paul

The research program INTERDYNAMIC aimed at a better understanding of climate dynamics using quantitative paleoclimate analyses in view of creating more reliable scenarios for future climate change. Between 2006 and 2013 a total of 22 projects were funded by the *Deutsche Forschungsgemeinschaft* (DFG) in this framework. The purpose of this book is to provide an overview of the key findings and a guide to the publications that resulted from each project.

INTERDYNAMIC was based on an integrated approach in paleoclimate research, in which all available paleoclimate archives (terrestrial and marine as well as ice cores) were combined in order to yield a comprehensive and quantitative analysis of global environmental variations. Moreover, through a close linkage between paleoclimate reconstructions and results from Earth-system models, detailed insights into the dynamics of climate variations have been gained, which are of relevance in assessing future climate changes.

The investigations in INTERDYNAMIC focused on the interglacials of the late Quaternary (including their onset and end). With respect to the global aspects of climate change, the focus in INTERDYNAMIC has been on global and continent- or basin-wide scales. Specifically, INTERDYNAMIC was directed towards the following overarching topics:

- Amplitudes of natural climate variations on timescales of several years to millennia
- Variations of patterns of climate variability in time and space

---

M. Schulz (✉) · A. Paul
MARUM—Center for Marine Environmental Sciences and Faculty of Geosciences, University of Bremen, Bremen, Germany
e-mail: mschulz@marum.de

A. Paul
e-mail: apaul@marum.de

© The Author(s) 2015   1
M. Schulz and A. Paul (eds.), *Integrated Analysis of Interglacial Climate Dynamics (INTERDYNAMIC)*, SpringerBriefs in Earth System Sciences, DOI 10.1007/978-3-319-00693-2_1

- Evidence for abrupt changes in the large-scale circulation of the Atlantic Ocean in interglacials
- Control of biogeochemical feedback mechanisms on natural limits of atmospheric concentrations of greenhouse gases
- Linkages between climate and pre-industrial cultures

These topics were addressed through a combination of climate information from paleoclimate archives with Earth-system modeling. The approach included state-of-the-art proxies and a wide spectrum of models, ranging from Earth-system models of intermediate complexity to comprehensive general circulation models.

INTERDYNAMIC consisted exclusively of collaborative, so-called Dual$^+$ projects, in which at least two of the research fields: marine archives, terrestrial archives, ice cores and Earth-system modeling were represented. Key findings from each project are summarized in the individual chapters. These are organized in terms of going from longer to shorter timescales, and from global and hemispheric to regional scales.

The first chapters compare the response to climate forcings from orbital variations and changes in atmospheric greenhouse-gas concentrations for different interglacials. Kleinen et al. (Chap. Comparison of Climate and Carbon Cycle Dynamics During Late Quaternary Interglacials) compared climate and carbon cycle dynamics during several interglacials of the late Quaternary with an emphasis of understanding the balance between terrestrial and oceanic sources and sinks during the course of the Holocene, the last interglacial (LIG), and Marine Isotope Stage (MIS) 11. Based on a global synthesis of sea-surface temperature (SST), Milker et al. (Chap. Global Synthesis of Sea-Surface Temperature Trends During Marine Isotope Stage 11) demonstrated a detectable signature of orbital forcing in the global SST pattern during MIS 11. Orbitally-driven trends in SST were investigated by Lohmann et al. (Chap. Evaluation of Eemian and Holocene Climate Trends: Combining Marine Archives with Climate Modelling) for the Holocene and LIG. Their model-data comparison indicates a shift in the seasonal imprint on proxy reconstructions that needs to be taken into account when interpreting SST reconstructions. Statistical analyses of globally distributed SST records by Mudelsee and Lohmann (Chap. Climate Sensitivity During and Between Interglacials) provided evidence that climate sensitivity during the LIG and MIS 11 were not different from the Holocene.

Onset and end of an interglacial provide valuable opportunities to test our understanding of feedback mechanisms in the climate system. It has been suggested that low-latitude warming leads ice melting in the northern hemisphere by several thousand years during the onset of interglacials. This hypothesis is challenged by Eisenhauer et al. (Chap. Phase-Shift Between Surface Ocean Warming, Evaporation and Changes of Continental Ice Volume During Termination I Observed at Tropical Ocean Sediment Cores), who identified a significant bias in low-latitude SST reconstructions that may account for the apparent offset in time. Meltwater routed through the Gulf of Mexico into the North Atlantic is thought to affect the meridional overturning circulation during glacial-to-interglacial transitions. Nürnberg et al. (Chap. Loop Current Variability—Its Relation to Meridional Overturning

Circulation and the Impact of Mississippi Discharge) question the impact of mega-discharges from the Mississippi River on the large-scale overturning circulation. Prange et al. (Chap. Hydroclimatic Variability in the Panama Bight Region During Termination 1 and the Holocene) investigated the role of atmospheric vapor transport from the Atlantic to the Pacific Ocean across Central America during the last glacial termination and find that the feedback between cross-isthmus vapor flux and the strength of the Atlantic meridional overturning circulation is negligible. Govin et al. (Chap. What Ends an Interglacial? Feedbacks Between Tropical Rainfall, Atlantic Climate and Ice Sheets During the Last Interglacial) investigated the chain of events that accompany the end of an interglacial. For the end of the LIG they find substantial shifts in the South American hydrologic cycle and upper tropical Atlantic salinities that may have affected the large-scale circulation of the Atlantic Ocean and North Atlantic climate.

Several projects addressed amplitudes of natural climate variations and changes in patterns of climate variability from a hemispheric or basin-wide perspective. Paleoceanographic reconstructions in the gateway between the Atlantic and Arctic Oceans by Spielhagen et al. (Chap. Holocene Environmental Variability in the Arctic Gateway) revealed highly variable sea-ice conditions during the Holocene. A long-term cooling trend was rapidly reversed 100 years ago and replaced by general warming in the Arctic. Morley et al. (Chap. Detecting Holocene Changes in the Atlantic Meridional Overturning Circulation: Integration of Proxy Data and Climate Simulations) addressed how climate variability at multidecadal-to-century timescales is communicated between high and low latitudes. They demonstrated an important role of central-water circulation in the North Atlantic Ocean to transfer regional climate signatures of various forcings (freshwater flux, solar variability, orbital parameters) to a hemispheric or global scale. On multidecadal and millennial timescales, precipitation variability in the Caribbean region during the Holocene was strongly linked to SST changes in the North Atlantic Ocean, namely the Atlantic Multidecadal Oscillation and variations in the strength of the Atlantic Meridional Overturning Circulation (Felis et al., Chap. Control of Seasonality and Interannual to Centennial Climate Variability in the Caribbean During the Holocene—Combining Coral Records, Stalagmite Records and Climate Models). For the westerlies in the southern hemisphere Lamy et al. (Chap. The Southern Westerlies During the Holocene: Paleoenvironmental Reconstructions from Chilean Lake, Fjord, and Ocean Margin Sediments Combined with Climate Modeling) revealed a distinct latitudinal anti-phasing of wind changes between the core and northern margin of the wind belt over the Holocene on centennial-to-millennial timescales. Changes in atmospheric transport were also found to be relevant by Wegner et al. (Chap. Mineral Dust Variability in Antarctic Ice for Different Climate Conditions) to account for variations in dust transport to Antarctica on glacial-to-interglacial timescales.

Changes in the hydrological cycle have been studied in several projects across a range of timescales and regions. For the Indian monsoon system, Schneider et al. (Chap. Model-Data Synthesis of Monsoon Amplitudes for the Holocene and Eemian) reconstructed similar conditions for the Holocene and the LIG, whereas climate-model simulations indicate a more intense hydrological cycle for the LIG.

For the Asian monsoon system, Dallmeyer et al. (Chap. Vegetation, Climate, Man —Holocene Variability in Monsoonal Central Asia) found that the atmospheric response to Holocene insolation forcing was strongly modified by ocean-atmosphere interactions, while the interaction between vegetation and atmosphere had only minor influence on the large-scale Holocene climate change and was only important at a regional level. Hydrological changes in northwest Africa during the Holocene were studied by Schefuß et al. (Chap. North-West African Hydrologic Changes in the Holocene: A Combined Isotopic Data and Model Approach), who found no evidence for an abrupt change at the end of the African humid period suggesting a gradual precipitation decline. Linkages between climate and pre-industrial cultures were evaluated by Lemmen et al. (Chap. Global Land Use and Technological Evolution Simulations to Quantify Interactions Between Climate and Pre-industrial Cultures) for the Holocene transition to agriculture in western Eurasia. It was shown that migration is not a necessary prerequisite for this transition and that climate variability and extreme events had no significant impact, which reflects societal resilience.

Several regional studies were carried out in the framework of INTERDYNAMIC. Interactions between climate variations, biogeochemical cycles, and ecosystem variability have been studied in the Eastern Mediterranean Sea during the formation of Sapropel S1 in the Holocene. Schmiedl et al. (Chap. Holocene Climate Dynamics, Biogeochemical Cycles and Ecosystem Variability in the Eastern Mediterranean Sea) provide evidence for a scenario, in which sufficient organic matter for sapropel formation was buried under oligotrophic conditions in an anoxic water column, hence refuting the "high-productivity" hypothesis. Environmental and climate changes during the last two glacial terminations and inter-glacials have been studied by Arz et al. (Chap. Environmental and Climate Dynamics During the Last Two Glacial Terminations and Interglacials in the Black Sea/Northern Anatolian Region) in the Black Sea and northern Anatolian region. Holocene and LIG developed differently, with warmer and moister conditions prevailing during the LIG. Furthermore, major fluctuations in the hydrological state of the Black Sea were closely linked to changes in the terrestrial environment. Reconstructions of summer precipitation variability and flood events for the Main region in southern Germany by Schoenbein et al. (Chap. Seasonal Reconstruction of Summer Precipitation Variability and Dating of Flood Events for the Millennium Between 3250 and 2250 Years BC for the Main Region, Southern Germany) revealed a noticeable excursion towards drier conditions around 2750 BC. In addition, a period of high flood frequency from 2991 BC to 2693 BC has been identified. The evolution of precipitation and its variability over Europe and the Mediterranean over the last two millennia was investigated by Gomez-Navarro et al. (Chap. Precipitation in the Past Millennium in Europe—Extension to Roman Times), who demonstrated the added value of regional climate models for downscaling. As a result, the Medieval Climate Anomaly was characterized by periods with warmer and drier summer conditions, and the Little Ice Age was characterized by periods with colder and wetter summer conditions, respectively.

In addition to the scientific achievements, INTERDYNAMIC provided a platform that facilitated the exchange of knowledge in the German paleoclimate community and linked their activities with international programs and projects. By training a large number of young scientists, the program has also helped to create a new generation of paleoclimatologists with a great awareness of adjacent disciplines. To this end, the Dual$^+$ concept of combining different expertise in a project right from its start has proven very useful. Specifically, the exchange between scientists reconstructing past climate changes and scientists numerically modeling past climate changes has benefitted greatly from INTERDYNAMIC.

Acknowledgements We thank the DFG for generously supporting INTERDYNMIC over a 6-year period. We are grateful to the anonymous mid-term reviewers, who helped to shape the program.

Comparison of Climate and Carbon Cycle Dynamics During Late Quaternary Interglacials

Thomas Kleinen, Elena Bezrukova, Victor Brovkin,
Hubertus Fischer, Steffi Hildebrandt, Stefanie Müller,
Matthias Prange, Rima Rachmayani, Jochen Schmitt,
Robert Schneider, Michael Schulz and Pavel Tarasov

Abstract Within the project COIN we investigated climate and carbon cycle changes during late Quaternary interglacials using ice core and terrestrial archives, as well as earth system models. The Holocene carbon cycle dynamics can be explained both in models and data by natural forcings, where the increase in CO_2 is due to oceanic carbon release, while the land is a carbon sink. Climate changes during MIS 11.3 were mainly driven by insolation changes, showing substantial differences within the interglacial. Terrestrial reconstructions and model results agree, though data coverage leaves room for improvement. The carbon cycle dynamics during MIS 11.3 can generally be explained by the same forcing mechanisms as for the Holocene, while model and data disagree during MIS 5.5, showing an increasing CO_2 trend in the model though reconstructions are constant.

Keywords Interglacial · Carbon cycle · $\delta^{13}CO_2$ · Holocene · MIS 11.3 · MIS 5.5 · Ice core data · Terrestrial data · Earth system model · CO_2

T. Kleinen (✉) · V. Brovkin
Max Planck Institute for Meteorology, Hamburg, Germany
e-mail: thomas.kleinen@mpimet.mpg.de

E. Bezrukova
A.P. Vinogradov Institute of Geochemistry SB RAS, Irkutsk, Russia

H. Fischer · J. Schmitt · R. Schneider
Climate and Environmental Physics, Physics Institute and Oeschger Centre for Climate Change Research, University of Bern, Bern, Switzerland

S. Hildebrandt · S. Müller · P. Tarasov
Institute of Geological Sciences, Freie Universität Berlin, Berlin, Germany

M. Prange · R. Rachmayani · M. Schulz
MARUM—Center for Marine Environmental Sciences and Faculty of Geosciences, University of Bremen, Bremen, Germany

© The Author(s) 2015
M. Schulz and A. Paul (eds.), *Integrated Analysis of Interglacial Climate Dynamics (INTERDYNAMIC)*, SpringerBriefs in Earth System Sciences, DOI 10.1007/978-3-319-00693-2_2

1 Introduction

The successful simulation of past climate changes is an important indicator of the ability of climate models to forecast future climate changes. While the climate of the Holocene has been relatively well investigated with global climate models, previous interglacials received much less attention.

Within this project, we provided quantitative reconstructions of several interglacials using ice core and terrestrial archives on the one hand and a hierarchy of Earth System models on the other hand. While components such as peat accumulation and $CaCO_3$ sedimentation, which are necessary to explain the ice core data, were previously missing from carbon cycle models, the ice core community only recently succeeded in measuring carbon isotopic data, which provides important constraints on the mechanisms of the CO_2 changes observed.

2 Materials and Methods

2.1 CLIMBER2-LPJ

CLIMBER2-LPJ (Kleinen et al. 2010) is a coupled climate carbon cycle model consisting of the earth system model of intermediate complexity (EMIC) CLIM-BER2 coupled to the dynamic global vegetation model (DGVM) LPJ, extended by a model of peat accumulation and decay (Kleinen et al. 2012).

2.2 Community Climate System Model Version 3 (CCSM3)

The National Center for Atmospheric Research (NCAR) CCSM3 is a state-of-the-art coupled climate model (Yeager et al. 2006). The resolution of the atmosphere is T31 (3.75° transform grid), while the ocean model has a horizontal resolution of 3° with a finer resolution around the equator.

2.3 Ice Core Measurements of Atmospheric $\delta^{13}CO_2$

The carbon isotopic composition of atmospheric CO_2 provides important benchmark data to test hypotheses on past changes in the global carbon cycle. Previous $\delta^{13}CO_2$ measurements were limited in resolution and precision, requiring large ice samples. We employed a novel sublimation extraction technique (Schmitt et al. 2011), cutting down sample size to 30 g of ice, while improving precision by a factor of up to 2–0.06 ‰. This has been applied to ice core samples from the

EPICA Dome C and Talos Dome ice cores, Antarctica, over the time interval 1 thousand years (ka) before present (BP)–25 ka BP and 105–155 ka BP (Schneider et al. 2013).

2.4 Palaeodata Assemblage

We reviewed a total of 162 publications affiliated with MIS 11.3 climate (Kleinen et al. 2014) and a set of 234 records for MIS 5.5. However, this extensive search revealed that the majority of the publications provide only qualitative, discontinuous and/or poorly dated information about the past climate. Therefore we focused on more recently published continuous records of interglacial climate and vegetation with well constrained chronologies, most suitable for a robust data-model comparison (e.g., Kleinen et al. 2011).

3 Key Findings

3.1 Trends in Interglacial Carbon Cycle Dynamics

During Termination I, atmospheric CO_2 rose as the ocean released carbon to the atmosphere (Schmitt et al. 2012). The $\delta^{13}CO_2$ measurements over Termination II point at the same processes being responsible for the CO_2 increase as in Termination I, however, with different phasing and magnitude (Schneider et al. 2013). After an initial CO_2 peak, CO_2 decreases during most interglacials, while the Holocene and MIS 11.3 reveal CO_2 increases by about 20 and 10 ppmv, respectively. Ruddiman (2003) interprets the rising CO_2 during the Holocene as the onset of the Anthropocene. However, simulations with CLIMBER2-LPJ suggest that the ocean mostly operates as a source of CO_2 to the atmosphere during interglacials (Kleinen et al. 2010). Carbonate compensation and the excessive accumulation of $CaCO_3$ in coral reefs lead to a slow CO_2 release to the atmosphere (Kleinen et al. 2010). The role of land carbon is more complex. During the Holocene, the land mainly serves as a sink of carbon since soil and peat storages on land are slowly growing (Kleinen et al. 2012). Reconstructed $\delta^{13}CO_2$ data from ice cores (Elsig et al. 2009; Schmitt et al. 2012; Schneider et al. 2013) suggests a continuous increase in the terrestrial carbon storage from 12 to 6 ka BP and a mostly neutral role of the land from 6 to 2 ka (Fig. 1a, b), invalidating the early Anthropocene hypothesis. The CLIMBER2-LPJ simulation for MIS 11.3 using the same forcing setup as for the Holocene shows CO_2 dynamics close to observations (Fig. 1d). Our new ice core measurements also show constant CO_2 concentrations over most of MIS 5.5. However, the CLIMBER2-LPJ results are in disagreement with CO_2 reconstructions (Fig. 1c), suggesting that the model still misses some important

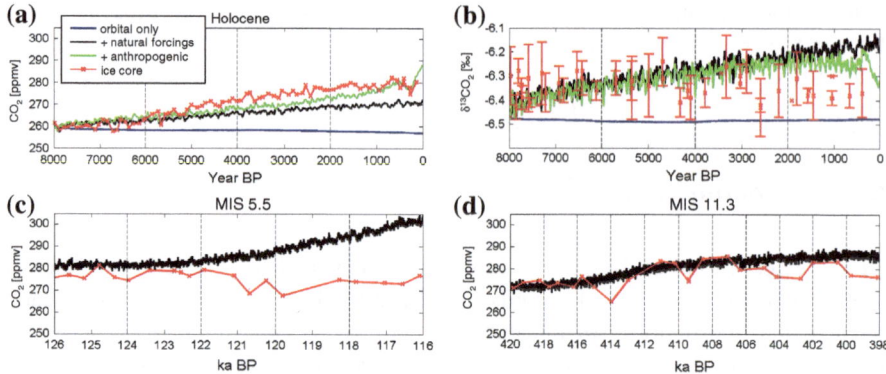

Fig. 1 CO_2 development during interglacials. **a** Holocene CO_2, **b** Holocene $\delta^{13}CO_2$, **c** MIS 5.5 CO_2, **d** MIS 11.3 CO_2. Ice core measurements (*red, crosses*), orbital forcing only (*blue*), plus natural forcings (*peat and corals, black*), plus anthropogenic landuse emissions (*green*). Ice core data as in Elsig et al. (2009), Schneider et al. (2013) and Petit et al. (1999) for Holocene, MIS 5.5, and MIS 11.3, respectively

process. We conclude that the difference between land and ocean biogeochemical processes during interglacials can drive atmospheric CO_2 either upward or downward, depending on the strength of warming controlled by orbital forcing and the history of carbon changes during the preceding terminations.

3.2 Climate and Vegetation Changes

Climate responds to numerous forcings. During interglacials, ice sheet changes are negligible, leading to a strong influence of insolation and greenhouse gas changes on climate. During the early Holocene, the CO_2 concentration (~ 265 ppm) was somewhat lower than preindustrial (280 ppm), but summer insolation in the high northern latitudes was substantially higher. This led to a strong summer warming and major changes in vegetation, a northerly advance of the northern tree line. Changes in tree cover for CLIMBER2-LPJ and 8 ka BP are shown in Fig. 2a. These tree cover changes compare favourably with reconstructions of woody cover from terrestrial records as shown in Fig. 2b. Since we performed transient integrations of the climate model and used tree cover reconstructions at high temporal resolution, we could show that the changes in tree cover occurred at similar times in model and reconstructions, though locally deviations of up to 1,000 years occurred (Kleinen et al. 2011).

For MIS 11.3 (Kleinen et al. 2014), we performed transient experiments with CLIMBER2-LPJ and time slice experiments with CCSM3, which were used to drive the LPJ DGVM. Climate changes are surprisingly similar for the two models, though CCSM3 reacts somewhat more strongly to insolation changes. Model results show large variations for MIS 11.3 climate, with European summer

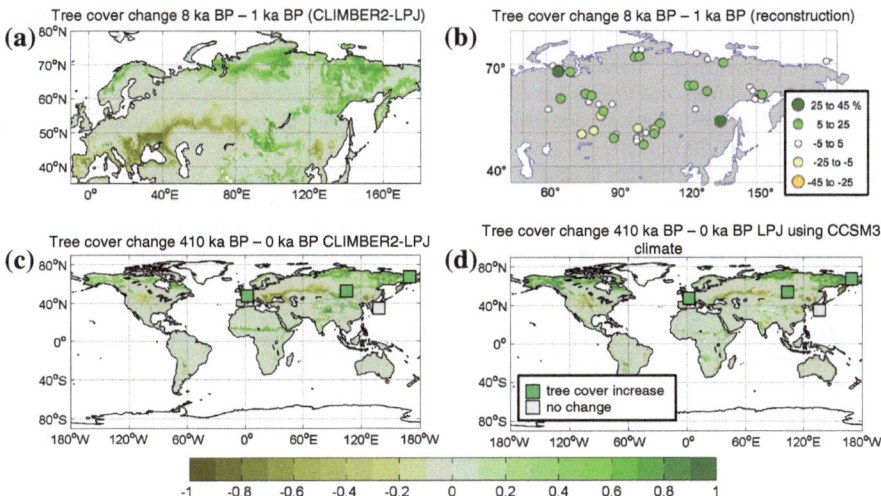

Fig. 2 Modeled and reconstructed tree cover change during the Holocene and MIS 11.3. **a** 8–1 ka BP, CLIMBER2-LPJ; **b** 8–1 ka BP, reconstruction (after Kleinen et al. 2011); **c** 410–0 ka BP, CLIMBER2-LPJ; **d** 410–0 ka BP, LPJ using CCSM3 climate. Reconstructed tree cover changes for MIS 11.3 are shown as *squares* in **c** and **d** (after Kleinen et al. 2014)

temperatures, for example, ranging from 2 °C colder to 3 °C warmer than preindustrial. For vegetation changes, shown in Fig. 2c, d for 410 ka BP, the results from the two models once more agree well, showing a northward shift of the taiga-tundra boundary in the high northern latitudes. Reconstructions from terrestrial records (Kleinen et al. 2014) show a similar magnitude of climate change, but the coverage by well-dated high-resolution records is very limited, preventing a model-data comparison for more than a few points.

References

Elsig J, Schmitt J, Leuenberger D, Schneider R, Eyer M, Leuenberger M, Joos F, Fischer H, Stocker TF (2009) Stable isotope constraints on Holocene carbon cycle changes from an Antarctic ice core. Nature 461:507–510

Kleinen T, Brovkin V, von Bloh W, Archer D, Munhoven G (2010) Holocene carbon cycle dynamics. Geophys Res Lett 37:L02705. doi:10.1029/2009GL041391

Kleinen T, Tarasov P, Brovkin V, Andreev A, Stebich M (2011) Comparison of modeled and reconstructed changes in forest cover through the past 8000 years: Eurasian perspective. Holocene 21(5):723–734

Kleinen T, Brovkin V, Schuldt RJ (2012) A dynamic model of wetland extent and peat accumulation: results for the Holocene. Biogeosciences 9:235–248

Kleinen T, Hildebrandt S, Prange M, Rachmayani R, Müller S, Bezrukova E, Brovkin V, Tarasov P (2014) The climate and vegetation of marine isotope stage 11—model results and proxy-based reconstructions at global and regional scale. Quatern Int. doi:10.1016/j.quaint.2013.12.028

Petit JR, Jouzel J, Raynaud D, Barkov NI, Barnola JM, Basile I, Benders M, Chappellaz J, Davis M, Delayque G, Delmotte M, Kotlyakov VM, Legrand M, Lipenkov VY, Lorius C, Pépin L, Ritz C, Saltzman E, Stievenard M (1999) Climate and atmospheric history of the past 420,000 years from the Vostok ice core, Antarctica. Nature 399:429–436

Ruddiman WF (2003) The anthropogenic Greenhouse Era began thousands of years ago. Clim Change 61:261–293

Schmitt J, Schneider R, Fischer H (2011) A sublimation technique for high-precision $\delta^{13}C$ on CO_2 and CO_2 mixing ratio from air trapped in deep ice cores. Atmos Meas Tech 4:1445–1461

Schmitt J, Schneider R, Elsig J, Leuenberger D, Lourantou A, Chappellaz J, Köhler P, Joos F, Stocker TF, Leuenberger M, Fischer H (2012) Carbon isotope constraints on the deglacial CO_2 rise from ice cores. Science 336:711–714

Schneider R, Schmitt J, Köhler P, Joos F, Fischer H (2013) A reconstruction of atmospheric carbon dioxide and its stable carbon isotopic composition from the penultimate glacial maximum to the glacial inception. Clim Past 9:2507–2523

Yeager SG, Shields CA, Large WG, Hack JJ (2006) The low-resolution CCSM3. J Clim 19:2545–2566

Global Synthesis of Sea-Surface Temperature Trends During Marine Isotope Stage 11

Yvonne Milker, Rima Rachmayani, Manuel F.G. Weinkauf,
Matthias Prange, Markus Raitzsch, Michael Schulz
and Michal Kučera

Abstract To examine the sea-surface temperature (SST) evolution during inter-glacial Marine Isotope Stage (MIS) 11, we compiled a database of 78 SST records from 57 sites. We aligned these records by oxygen-isotope stratigraphy and sub-jected them to an Empirical Orthogonal Function (EOF) analysis. The principal SST trend (EOF1) reflects a rapid deglacial warming of the surface ocean in pace with carbon dioxide rise during Termination V, followed by a broad SST optimum centered at $\sim$410 thousand years (ka) before present (BP). The second EOF indicates the existence of a regional SST trend, characterized by a delayed onset of the SST optimum, followed by a prolonged period of warmer temperatures. The proxy-based SST patterns were compared to CCSM3 climate model runs for three time slices representing different orbital configurations during MIS 11. Although the modeled SST anomalies are characterized by generally lower variance, corre-lation between modeled and reconstructed SST anomalies suggests a detectable signature of astronomical forcing in MIS 11 climate trends.

Keywords MIS 11 · Interglacial · Quaternary · Sea-surface temperature · Data-model comparison · Empirical orthogonal function

Y. Milker (✉) · M.F.G. Weinkauf · M. Kučera
Department of Geosciences, University of Tübingen, Tübingen, Germany
e-mail: Yvonne.Milker@uni-hamburg.de

Y. Milker · R. Rachmayani · M.F.G. Weinkauf · M. Prange · M. Schulz · M. Kučera
MARUM—Center for Marine Environmental Sciences and Faculty of Geosciences,
University of Bremen, Bremen, Germany

M. Raitzsch
Helmholtz Centre for Polar and Marine Research, Alfred Wegener Institute,
Bremerhaven, Germany

Y. Milker
Center for Earth System Research and Sustainability (CEN), Institute for Geology,
University of Hamburg, Hamburg, Germany

© The Author(s) 2015
M. Schulz and A. Paul (eds.), *Integrated Analysis of Interglacial Climate Dynamics (INTERDYNAMIC)*, SpringerBriefs in Earth System Sciences,
DOI 10.1007/978-3-319-00693-2_3

1 Introduction

Marine Isotope Stage (MIS) 11 (424–374 thousand years (ka) before present (BP)) (Lisiecki and Raymo 2005) was characterized by an unusually long climatic optimum (Tzedakis et al. 2009) with warm conditions lasting longer than in any other mid to late Pleistocene interglacial (Jouzel et al. 2007). The orbital forcing during MIS 11 was characterized by a different phasing of obliquity and precession leading to lower amplitude insolation cycles compared to the Holocene (Loutre and Berger 2003). In contrast to the differences in orbital parameters, the greenhouse gas concentrations in the atmosphere during MIS 11 were similar to the preindustrial Holocene (Siegenthaler et al. 2005). Until now, the coherency of sea-surface temperature (SST) trends during MIS 11 has never been assessed on a global basis. Such comparison is essential for understanding the relationship between SST and climate forcing. In this project, a global compilation of SST records for MIS 11 has been produced, including records covering a large portion of the world ocean on both hemispheres. Our aim was to analyze the roles of orbital and greenhouse gas forcing in MIS 11 climate variability, to detect regional climate variability reflected in SST trends, and to compare the temporal-spatial climate variability simulated by a state-of-the-art climate model for orbital configuration extremes of MIS 11 with that found in proxy records in order to answer the overall question: What are the amplitudes of natural climate variations on timescales of several years to millennia and how do patterns of climate variability vary in time and space?

2 Materials and Methods

We compiled a total of 78 marine SST records from 57 sites (Fig. 1a). We have only used SST records for which also stable oxygen isotope data are available with a sufficient temporal resolution to establish a robust stratigraphic framework. The stratigraphic tuning was carried out against the LR04 stack (Lisiecki and Raymo 2005) [see Milker et al. (2013) for further information]. To extract principal SST modes, the tuned SST records were standardized and interpolated linearly to a temporal resolution of 1,000 years and subjected to an Empirical Orthogonal Function (EOF) analysis. To examine the robustness of the EOF results, we repeated the analysis on randomly resampled age models and SST values with increasing uncertainty. We further performed a resampling approach to test the sensitivity of the EOFs to the number of records used in the analysis. Finally, we compared SST anomalies in the proxy data with climate model outputs for three time slices (394, 405 and 416 ka BP) representing different extremes of the orbital configuration during MIS 11. The time slice simulations were performed with CCSM3, the National Center for Atmospheric Research (NCAR) Community Climate System Model version 3 (see Chap. Comparison of Climate and Carbon Cycle Dynamics During Late Quaternary Interglacials).

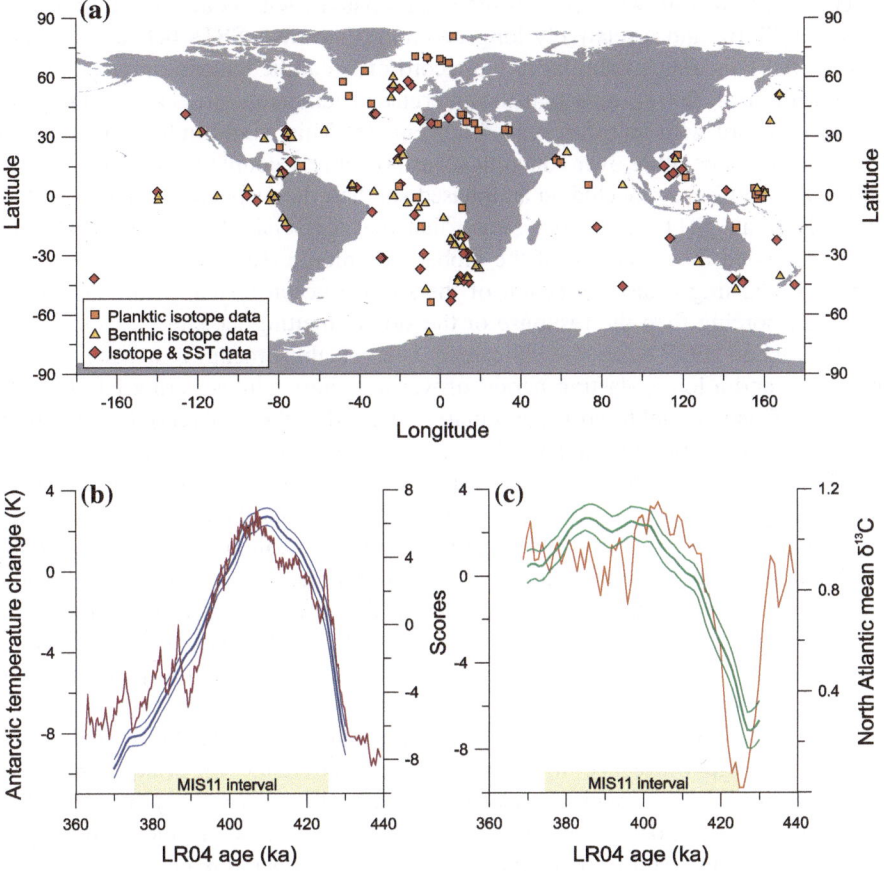

Fig. 1 a The location of stable isotope and SST data compiled in the project. **b** Scores of the first EOF with the 95 % confidence interval (age uncertainty set to 5 ka and temperature uncertainty set to 1 °C) compared to the Antarctic temperature change during MIS 11 (Jouzel et al. 2007), **c** Scores of the second EOF with the 95 % confidence interval (age uncertainty set to 5 ka and temperature uncertainty set to 1 °C) compared to deep water mean $\delta^{13}C$ values of the North Atlantic used as a proxy for North Atlantic Deepwater (NADW) production (see also Lisiecki et al. 2008)

3 Key Findings

The EOF analysis reveals two main SST modes (Fig. 1b, c), which explain nearly 70 % of the variation in the dataset, with 49 % explained by EOF1 and ~18 % explained by EOF2. Although we found a stronger influence of temperature uncertainty on the EOF robustness compared to the uncertainty of the age model and reduction of number of records included into analysis, both the shape of the first two EOFs and the amount of variance explained by them are remarkably robust to age-model and temperature uncertainties (Milker et al. 2013).

The first EOF follows a glacial-interglacial pattern with cold SSTs during MIS 12 and MIS 10, and a relatively long duration of warmer SSTs between 416 and 405 ka BP. This trend is similar to the record of Antarctic temperature during MIS 11 (Jouzel et al. 2007), although the interglacial temperature peak is leading the Antarctic record with an offset of ~4 ka (Fig. 1b).This would indicate that during MIS 11, the temperature over Antarctica was not closely coupled to the global mean SST and might have reflected an antiphased southern hemisphere insolation pattern (Laepple et al. 2011). It further seems that the deglacial SST rise, indicated by EOF1 preceded the reduction of the global ice volume (Elderfield et al. 2012) by ~5 ka suggesting a faster reaction of the surface ocean to insolation and greenhouse gas forcing than the response of the slowly melting ice sheets.

The second EOF's scores indicate a later establishment of a relative SST maximum and a longer-lasting period of warmer temperatures during late MIS 11 (Fig. 1c). This regional trend is particularly reflected in the SST records of the mid-latitude North Atlantic Ocean and Mediterranean Sea. The apparently later onset of the MIS 11 optimum and the longer duration of interglacial warmth have also been observed by Voelker et al. (2010), who hypothesized that the associated sustained meltwater input to the (sub-)polar regions may have resulted in a weaker Altantic Meridional Overturning circulation (AMOC). Similarly, mean $\delta^{13}C$ of benthic foraminifera from deeper waters in the North Atlantic used as a proxy for North Atlantic Deepwater (NADW) production (Lisiecki et al. 2008) show a trend of increasing NADW production between 410 and 400 ka BP which is quite similar to the EOF2 scores (Fig. 1c).

The comparison of the proxy-based SST anomalies with CCSM3 model results revealed a large difference in their variance. The range of proxy-based SST anomalies is ~4 °C, whereas modeled SST anomalies vary rarely by more than 1 °C (Milker et al. 2013) (Fig. 2a). The much lower variance in modeled temperature trends might result from an underestimation of temperature changes in climate models, an overestimated proxy SST variability, or from a combination of both. Underestimation of climate variability in model simulations may be caused by shortcomings in the model physics and/or missing climate components resulting in a lack of potentially important feedback mechanisms. Higher variance in the proxy data may result from noise and calibration uncertainties, from larger shifts in the ecology of the microfossils or changes in seasonality and vertical habitats. Despite the large differences in variance and considering all the potential sources of uncertainty in the proxy-based SST values, it is remarkable that in several cases a visual agreement between the direction of proxy and model SST changes emerged (Fig. 2b). Moreover for the boreal summer season of the 416 ka BP time slice as well as for the boreal winter season of the 405 ka BP time slice, a statistically significant correlation between the proxy-based and modeled SST anomalies was found (Milker et al. 2013).This indicates that orbital forcing, the major driver in the CCSM3 experiments, has left a detectable signature in the global SST pattern during MIS 11, despite its unusually low amplitude.

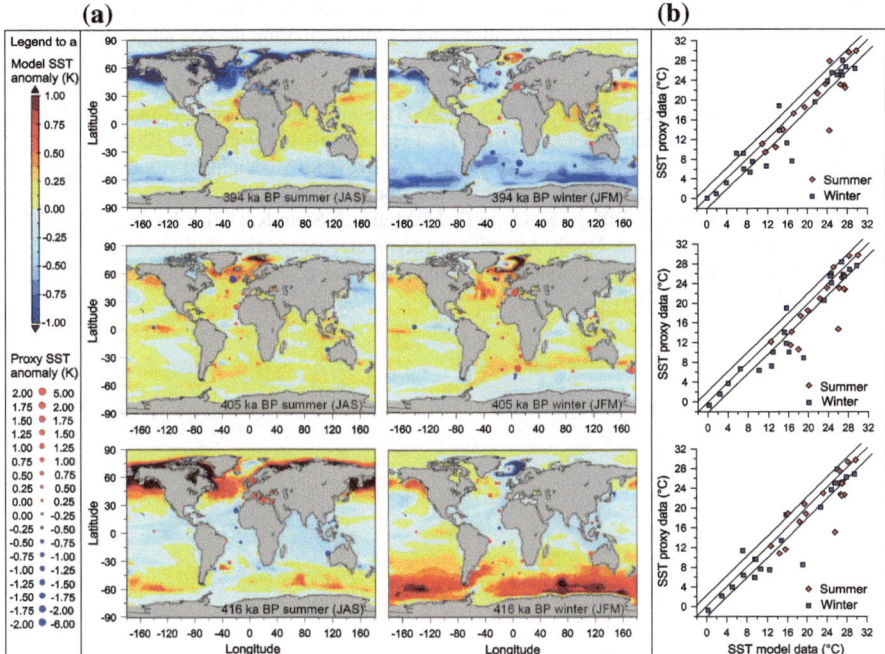

Fig. 2 The distribution of summer and winter SST anomalies modeled by CCSM3 versus proxy-based SST anomalies for the three selected time slices **a** and absolute modeled SST values versus absolute proxy SST values for the same times slices **b**. The *dashed lines* in **b** highlight ±2 °C temperature uncertainty intervals

References

Elderfield H, Ferretti P, Greaves M, Crowhurst S, McCave IN, Hodell D, Piotrowski AM (2012) Evolution of ocean temperature and ice volume through the Mid-Pleistocene climate transition. Science 337:704–709. doi:10.1126/science.1221294

Jouzel J, Masson-Delmotte V, Cattani O, Dreyfus G, Falourd S, Hoffmann G, Minster B, Nouet J, Barnola JM, Chappellaz J, Fischer H, Gallet JC, Johnsen S, Leuenberger M, Loulergue L, Luethi D, Oerter H, Parrenin F, Raisbeck G, Raynaud D, Schilt A, Schwander J, Selmo E, Souchez R, Spahni R, Stauffer B, Steffensen JP, Stenni B, Stocker TF, Tison JL, Werner M, Wolff EW (2007) Orbital and millennial Antarctic climate variability over the past 800,000 years. Science 317:793–796. doi:10.1126/science.1141038

Laepple T, Werner M, Lohmann G (2011) Synchronicity of Antarctic temperatures and local solar insolation on orbital timescales. Nature 471:91–94. doi:10.1038/nature09825

Lisiecki LE, Raymo ME (2005) A Pliocene-Pleistocene stack of 57 globally distributed benthic δ^{18}O records. Paleoceanography 20:PA1003. doi:10.1029/2004PA001071

Lisiecki LE, Raymo ME, Curry WB (2008) Atlantic overturning responses to late Pleistocene climate forcings. Nature 456:85–88. doi:10.1038/nature07425

Loutre MF, Berger A (2003) Marine isotope stage 11 as an analogue for the present interglacial. Global Planet Change 36:209–217. doi:10.1016/S0921-8181(02)00186-8

Milker Y, Rachmayani R, Weinkauf M, Prange M, Raitzsch M, Schulz M, Kučera M (2013) Global and regional sea surface temperature trends during marine isotope stage 11. Clim Past Discuss 9:837–890. doi:10.5194/cpd-9-837-2013

Siegenthaler U, Stocker TF, Monnin E, Lüthi D, Schwander J, Stauffer B, Raynaud D, Bamola JM, Fischer H, Masson-Delmotte V, Jouzel J (2005) Stable carbon cycle-climate relationship during the late Pleistocene. Science 310:1313–1317. doi:10.1126/science.1120130

Tzedakis PC, Raynaud D, McManus JF, Berger A, Brovkin V, Kiefer T (2009) Interglacial diversity. Nat Geosci 2:751–755. doi:10.1038/ngeo660

Voelker AHL, Rodrigues T, Billups K, Oppo D, McManus J, Stein R, Hefter J, Grimalt JO (2010) Variations in mid-latitude North Atlantic surface water properties during the mid-Brunhes (MIS 9–14) and their implications for the thermohaline circulation. Clim Past 6:531–552. doi:10.5194/cp-6-531-2010

Climate Sensitivity During and Between Interglacials

Manfred Mudelsee and Gerrit Lohmann

Abstract Studying the climate dynamics of past interglacials (IGs) may help to better assess the anthropogenically influenced dynamics of the current IG, the Holocene. We select IG sections from the longest ice core archive, EPICA Dome C (EDC), which covers the past 800 thousand years, and study as well several long, high-resolution marine sediment records. We analyze records of Antarctic temperature, radiative forcing (greenhouse gases and other factors), and sea-surface temperature (SST). Change-point regressions inform about longer-term climate changes and trends within IGs. Comparing trends in temperature with trends in forcing allows inference of longer-term IG climate sensitivities. Results from many records indicate deviations from a "Holocene climate optimum". IG sensitivities are found to be comparable to estimates for the instrumental period; warming or cooling phases during Marine Isotope Stage (MIS) 5 or 11 do not show significant differences in climate sensitivity.

Keywords Change points · Climate sensitivity · Greenhouse gases · Ice cores · Marine archives · Statistical time series analysis

1 Introduction

The effective climate sensitivity, S, is defined as the change in annual surface temperature, T, per change in radiative forcing, R. Knowledge about R is limited, which is reflected in an uncertain value of S. Published estimates for the time span since ~ 1850 range from ~ 0.3 to $1.5 \, \text{K} \, \text{W}^{-1} \, \text{m}^2$ [Masson-Delmotte et al. (2013) and references therein]. Analyses of temperature and forcing changes in the geologic

M. Mudelsee · G. Lohmann
Alfred Wegener Institute, Helmholtz Centre for Polar and Marine Research,
Bremerhaven, Germany

M. Mudelsee
Climate Risk Analysis, Heckenbeck, Bad Gandersheim, Germany

© The Author(s) 2015
M. Schulz and A. Paul (eds.), *Integrated Analysis of Interglacial Climate
Dynamics (INTERDYNAMIC)*, SpringerBriefs in Earth System Sciences,
DOI 10.1007/978-3-319-00693-2_4

past may increase the accuracy of sensitivity estimates owing to the possibly larger sizes of those changes (Köhler et al. 2010; Masson-Delmotte et al. 2013). This approach meets two obstacles. First, the paleo analyses regard longer timescales than from ~ 1850 to present, therefore additional feedback processes may act (Masson-Delmotte et al. 2013). Second, S may not be a constant but depend on the mean climatic state (Köhler et al. 2010; Masson-Delmotte et al. 2013).

To tap the potential of the paleoclimate archives (including model output) whilst bypassing both obstacles, the ClimSens project concentrated on the IG sections of various paleoclimatic records and fitted statistical change-point regression models to time-dependent temperature and forcing. This approach rests on the assumption of including in the analysis large enough changes in both variables while avoiding glacial intervals, which could exhibit different climate sensitivities. The ratio of the slopes (temperature over forcing) is used to estimate climate sensitivity on within-IG timescales. This method deviates from visually comparing both variables (Köhler et al. 2010) or from fitting an errors-in-variables regression to both variables (Mudelsee 2014a).

ClimSens contributes to INTERDYNAMIK via quantifying trends and studying feedback mechanisms also for older IGs (i.e., before MIS 1) utilizing data from ice cores, marine archives, and climate modeling.

2 Materials and Methods

The database (Mudelsee 2014b) comprises (1) one long [i.e., back to ~ 800 thousand years (ka) before present (BP)] radiative forcing anomaly time series (ΔR) based on EDC data (greenhouse gases) and modeling, (2) the classic long EDC Antarctic temperature series, (3) one long Northern Hemisphere (NH) surface-air temperature series based on $\delta^{18}O$ data and modeling, (4) two long low-resolution SST series, (5) five shorter (back up to ~ 400 ka BP) SST series, and (6) 21 high-resolution Holocene SST series. One wishes to have many records from distributed locations, on the other one needs a certain length and a high resolution to perform statistical analyses. The dates for the IG boundaries (Röthlisberger et al. 2008; NEEM Community Members 2013) are augmented by the date for the end of Marine Isotope Stage (MIS) 11, which was set to 395 ka BP (when EDC has the same temperature as at MIS 11 start 426 ka BP).

We fit a simple change-point regression model called the "break" (Mudelsee 2014a) to the records to infer changes in temperature and forcing. The model allows for the observed evidence that even within an IG, the climate needs neither be constant nor display a monotonic trend; rather, IG climate may show trend changes, as is seen for the Holocene and its "Holocene climate optimum" (Wanner et al. 2008). By systematically fitting the break model to a large set of records, we learn about changes in time, the signatures of "IG optima", and generally about the dynamics of IG climates. Despite that, the farther back in time, the fewer are the suitable records and the lower is the spatial coverage.

3 Key Findings

Results are shown for MIS 1, MIS 5, and MIS 11 (Fig. 1). In the search for IG optima, we distinguish qualitatively among different trend shapes. We define an optimum via a peak of high temperature or forcing, that means, an increasing trend in the earlier and a decreasing trend in the later part of an IG.

Nine out of 29 analyzed MIS 1 series exhibit an optimum. To quantify the timing of that Holocene climate optimum in SST records (note that ΔR, EDC, and NH do not peak in the Holocene), we first discard two results, where the apparent change point is at the interval bound. We conjecture that either measurement or proxy uncertainties produced a gross misestimation (MD01-2412) or that the size of the change is too small, which renders the timing estimate spurious (ODP 1089).

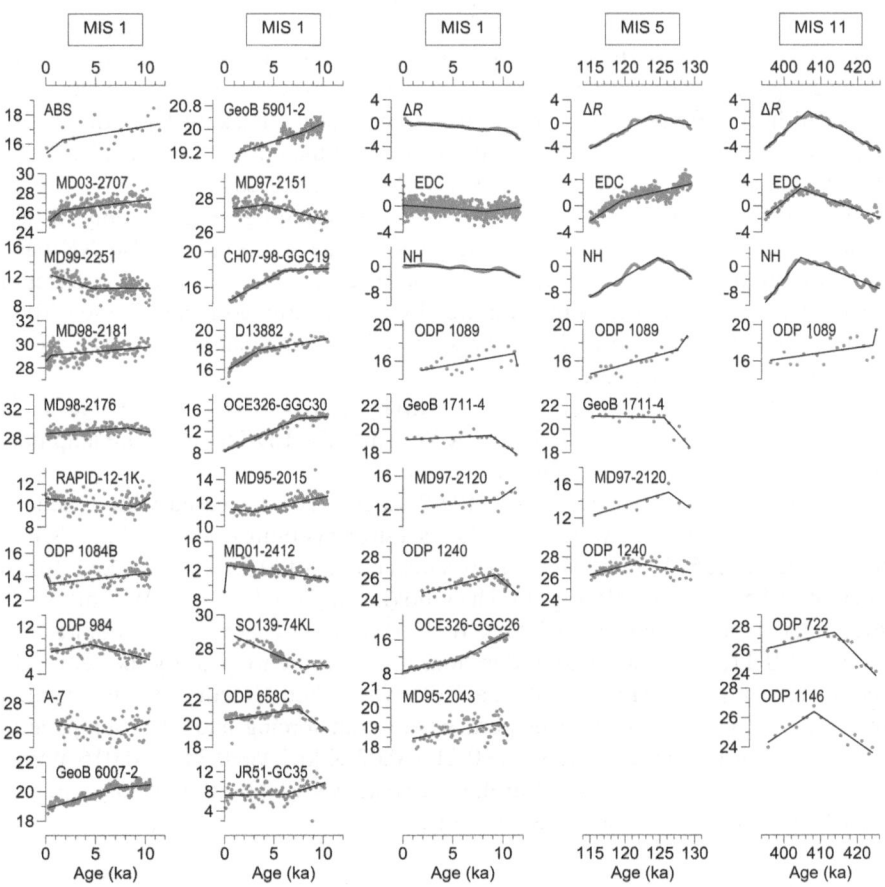

Fig. 1 Results. Panels show forcing and temperature time series (*gray dots*); units ΔR, W m^{-2}; all other, °C, all temperature records but EDC and NH measure SST

The results from the seven records left (MD98-2176, ODP 984, MD97-2151, ODP 658C, GeoB1711-4, ODP 1240 and MD95-2043) have a weighted average (Mudelsee 2014a) of the peak timing of 8.2 ka BP with an external (systematic) error of 0.6 ka and an internal (statistical) error of 0.2 ka. This value is more accurate than the assessment (~ 9 to 5 to 6 ka BP) in a review of Holocene climate (Wanner et al. 2008). The geographical pattern of the locations of the records (Mudelsee 2014b) that show this peak (Fig. 1) indicates a slight preference for northern latitudes, as Wanner et al. (2008) noted for the optimum, but also southern locations may display the peak (e.g., GeoB 1711-4). Wanner et al. (2008, p. 1792) noted that "orbital forcing (high summer insolation in the NH) was maximal at around 11,000 years BP [...], however, until about 9,000 years BP a large remnant ice sheet persisted in North America" with cooling effects. The relatively large systematic error, which is also larger than the statistical, hints at considerable variations among records in the timing. Also many records analyzed (Fig. 1) deviate from this simple "Holocene climate optimum" picture.

Four from seven analyzed MIS 5 series exhibit an optimum (Fig. 1). The sparseness of records, however, prohibits averaging and searching for a geographical pattern. The same applies also to MIS 11, where the IG peak behavior is observed in five from six data series. It may be that the peaking IG optimum is a real feature. It would be interesting to include into the change-point analyses also the SST datasets of a recent MIS 11 study (Milker et al. 2013).

The high-resolution ΔR and EDC time series, which agree remarkably in their trend shapes (Fig. 1), allow to estimate paleoclimate sensitivities. Also NH has a high resolution, but since ΔR was partly constructed using NH data (Köhler et al. 2010), the good agreement with ΔR in trend shape is partly artificial. The sensitivity estimate is obtained from dividing the slope of the temperature regression by the slope of the forcing regression.

For MIS 1 in the later part, the sensitivity is $S = (-0.11 \pm 0.31 \text{ K ka}^{-1})/(-0.14 \pm 0.01 \text{ W m}^{-2} \text{ ka}^{-1}) \approx 0.8 \pm 2.2 \text{ K W}^{-1} \text{ m}^2$. The EDC temperature amplitude is too small to allow a meaningful calculation.

For MIS 5 in the later part (since 123.7 ka BP), ΔR decreased with a slope of 0.64 W m^{-2} ka^{-1} (Fig. 1), while EDC shows a strong cooling of (0.70 ± 0.26) K ka^{-1} for the interval 115–119.5 ka BP and a weaker cooling of (0.25 ± 0.71) K ka^{-1} for the interval 119.5–129.5 ka BP (Fig. 1). The ratio (i.e., S) is (1.1 ± 0.4) K W^{-1} m^2 for the strong, later cooling and (0.4 ± 1.1) K W^{-1} m^2 for the weaker, earlier cooling.

For MIS 11, owing to its long duration, it is possible to quantify the sensitivity for its earlier part (before ~ 405 ka BP), where both forcing and temperature increased, as well as for its later part, where both forcing and EDC temperature decreased. For the earlier part, $S = (-0.21 \pm 0.02 \text{ K ka}^{-1})/(-0.349 \pm 0.016 \text{ W m}^{-2} \text{ ka}^{-1}) \approx 0.60 \pm 0.06 \text{ K W}^{-1} \text{ m}^2$. For the later part, $S = (0.44 \pm 0.05 \text{ K ka}^{-1})/(0.54 \pm 0.04 \text{ W m}^{-2} \text{ ka}^{-1}) \approx 0.8 \pm 0.1 \text{ K W}^{-1} \text{ m}^2$.

In summary, we find values from ΔR and EDC slopes that are comparable to values for the instrumental period. Errors are large due to sparseness and shortness of records. Warming or cooling phases during MIS 5 or MIS 11 do not show significant differences in climate sensitivity.

References

Köhler P, Bintanja R, Fischer H, Joos F, Knutti R, Lohmann G, Masson-Delmotte V (2010) What caused earth's temperature variations during the last 800,000 years? Data-based evidence on radiative forcing and constraints on climate sensitivity. Quat Sci Rev 29:129–145

Masson-Delmotte V, Schulz M, Abe-Ouchi A, Beer J, Ganopolski J, González Rouco JF, Jansen E, Lambeck K, Luterbacher J, Naish T, Osborn T, Otto-Bliesner B, Quinn T, Ramesh R, Rojas M, Shao X, Timmermann A (2013) Information from paleoclimate archives. In: Stocker TF, Qin D, Plattner GK, Tignor M, Allen SK, Boschung J, Nauels A, Xia Y, Bex V, Midgley PM (eds) Climate change 2013: the physical science basis. Contribution of working group I to the fifth assessment report of the intergovernmental panel on climate change. Cambridge, Cambridge University Press, pp 383–464

Milker Y, Rachmayani R, Weinkauf MFG, Prange M, Raitzsch M, Schulz M, Kučera M (2013) Global and regional sea surface temperature trends during marine isotope stage 11. Clim Past 9:2231–2252

Mudelsee M (2014a) Climate time series analysis: classical statistical and bootstrap methods, 2nd edn. Springer, Dordrecht

Mudelsee M (2014b) ClimSens project database (table 1), dataset #828671. doi:10.1594/PANGAEA.828671

NEEM Community Members (2013) Eemian interglacial reconstructed from a Greenland ice core. Nature 493:489–494

Röthlisberger R, Mudelsee M, Bigler M, de Angelis M, Fischer H, Hansson M, Lambert F, Masson-Delmotte V, Sime L, Udisti R, Wolff EW (2008) The southern hemisphere at glacial terminations: insights from the dome C ice core. Clim Past 4:345–356

Wanner H, Beer J, Bütikofer J, Crowley TJ, Cubasch U, Flückiger J, Goosse H, Grosjean M, Joos F, Kaplan JO, Küttel M, Müller SA, Prentice IC, Solomina O, Stocker TF, Tarasov P, Wagner M, Widmann M (2008) Mid- to late holocene climate change: an overview. Quat Sci Rev 27:1791–1828

What Ends an Interglacial? Feedbacks Between Tropical Rainfall, Atlantic Climate and Ice Sheets During the Last Interglacial

Aline Govin, Benjamin Blazey, Matthias Prange and André Paul

Abstract How long the present interglacial will last remains under debate. This project aims to determine the climatic mechanisms and sequence of events terminating an interglacial period. By comparing new paleoclimate reconstructions and climate model experiments, we investigate the impact of South American rainfall changes on tropical Atlantic sea-surface salinity and Atlantic thermohaline circulation at the end of the Last Interglacial (LIG). Model and proxy data show gradually intensifying South American monsoonal precipitation and enhanced Amazon discharge through the LIG, in response to increasing austral summer insolation. However, an increased meridional temperature gradient at the end of the LIG caused a strengthening of the North Brazil Current retroflection which deflected eastward the Amazon freshwater plume. Such changes in South American river discharge contributed to decrease tropical and North Atlantic surface salinities, resulting in a shift in regions of North Atlantic deep water convection and small reduction in deep water formation.

Keywords Last interglacial · Last glacial inception · Tropical precipitation · Atlantic ocean · Sea-surface salinity · Thermohaline circulation · Ice sheet · Model-data comparison

1 Introduction

When and how the present interglacial will end remains an open question (Tzedakis et al. 2012). With a relatively well-known climate, the Last Interglacial (LIG, 129 thousand years (ka) before present (BP)—116 ka BP) provides a unique framework to investigate the climatic mechanisms terminating an interglacial period.

A. Govin (✉) · B. Blazey · M. Prange · A. Paul
MARUM—Center for Marine Environmental Sciences and Faculty of Geosciences,
University of Bremen, Bremen, Germany
e-mail: agovin@marum.de

© The Author(s) 2015
M. Schulz and A. Paul (eds.), *Integrated Analysis of Interglacial Climate Dynamics (INTERDYNAMIC)*, SpringerBriefs in Earth System Sciences,
DOI 10.1007/978-3-319-00693-2_5

25

Although the decrease in boreal summer insolation is the primary driver ending the LIG, vegetation and ocean feedbacks are necessary to initiate ice sheet growth. However, the role of the prolonged North Atlantic warmth observed during the LIG is unclear: did it favor or delay the end of the LIG? Warm North Atlantic waters were likely sustained by an active thermohaline circulation (THC) across the LIG. Climate models, however, show no consistent temporal evolution of the THC throughout the LIG (Bakker et al. 2013), making involved climatic mechanisms uncertain. Past millennial-scale climate variability highlights strong interactions between tropical hydrology and the THC (Krebs and Timmermann 2007). However, the lack of detailed tropical proxy records and adequate simulations prevented investigating the role of feedbacks between tropical rainfall, Atlantic climate and ice sheets in the sequence of events ending the LIG.

By combining new high-resolution paleoclimate records and model experiments, we determine the influence of South American precipitation changes on tropical Atlantic salinity, North Atlantic mixed layer depth and the Atlantic THC. The effects of these changes on ice-sheet inception and growth at the end of the LIG presently remain under investigation.

2 Materials and Methods

We consider a transect of eight sediment cores from the South American margin (12°N–32°S). To reconstruct past changes in South American rainfall and freshwater delivery to the ocean, we applied well-established geochemical methods on the biogenic, terrigenous and organic fractions of the sediment. With the world's largest discharge, the Amazon River is the most likely river to impact tropical Atlantic salinity and the THC. We hence focus here on the northernmost sites (Fig. 1) located along the pathway of Amazon freshwater. We measured the sedimentary elemental composition which allows tracing the provenance of terrigenous material and reconstructing past climate variations over the source regions. We defined regional terrigenous endmembers based on six major elements and applied an endmember unmixing model to deduce the relative proportions of Amazonian Andean versus lowland material at 5°N and 9°N, and of Amazon versus Orinoco material at 12°N. See Govin et al. (2014) for detailed methods.

We use the Community Climate System Model version 3 (CCSM3) to co-verify the South American hydrologic reconstructions and investigate impacts on the climate system. The model was configured with T31 atmospheric resolution coupled to a nominally 3-degree ocean (Yeager et al. 2006). The fully coupled model further includes sea ice and a land surface model. We used CCSM3 for two equilibrated experiments, with orbital parameters and greenhouse gas concentrations at 125 and 115 ka BP as boundary conditions. The 125 ka BP experiment was initialized from a quasi-equilibrated 130 ka BP experiment. In turn, the 115 ka BP experiment was branched from the 125 ka BP experiment following an initial equilibration period. Both 115 and 125 ka BP simulations were allowed to

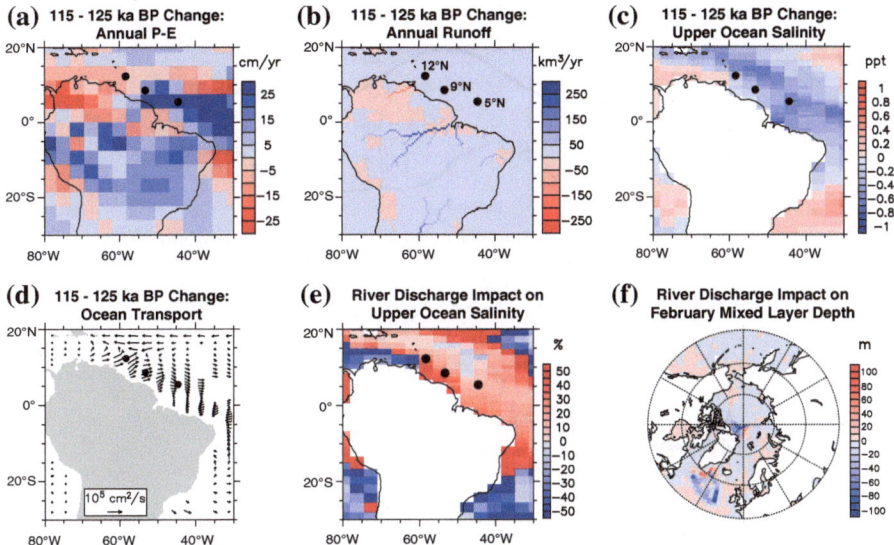

Fig. 1 Changes in South American climate from 125 to 115 ka BP (**a–d**). **a** Changes in total annual net precipitation (precipitation–evaporation), **b** changes in runoff, **c** change in upper ocean (0–150 m) salinity, **d** changes in water column integrated oceanic transport. Impact of 115–125 ka BP changes in South American river discharge (**e–f**). **e** Fraction of 115–125 ka BP upper ocean salinity change due to river runoff, **f** changes in February mixed layer depth in high northern latitudes linked to river discharge increase from 125 to 115 ka BP. *Black dots* show the locations of marine sediment cores (see caption of Fig. 2 for details)

equilibrate for 700 years. Two additional simulations were performed with 115 ka BP boundary conditions to isolate the impact of changes in river discharge, i.e. with South American river flow fixed to 115 or 125 ka BP conditions.

3 Key Findings

The sedimentary geochemical composition confirms that sites at 5°N and 9°N exclusively receive sediments from the Amazon River, while the Orinoco and Amazon Rivers both contribute to terrigenous input at 12°N (Govin et al. 2014). Cores at 5°N and 9°N exhibit a progressive increase in the relative proportion of Andean material between 126 and 111 ka BP (Fig. 2). It reflects the increasing input of detrital particles from Andean regions, where most of Amazon terrigenous material originates (Guyot et al. 2007), and agrees with the higher precipitation and runoff simulated over most of the Amazon basin at 115 ka BP compared to 125 ka BP (Fig. 1a, b). This result also agrees with western Amazonian speleothem records that suggest strong coupling between the intensity of the South American monsoon and

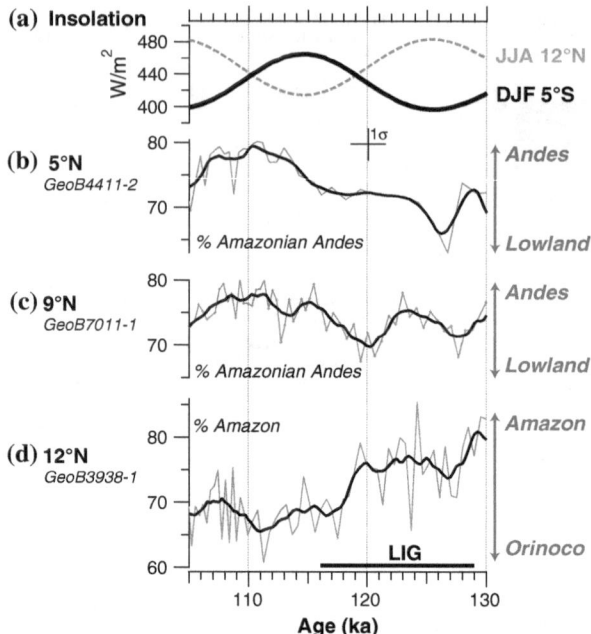

Fig. 2 Paleoclimatic records across the last interglacial (*LIG*). **a** Austral summer (DJF at 5°S, *black line*) and boreal summer (JJA at 12°N, *grey dotted line*) insolation, **b** proportion of Amazonian Andean (vs. lowland) material within the terrigenous fraction of core GeoB4411-2 (5.4°N, 44.5°W, 3,295 m), **c** same as **b** for core GeoB7011-1 (8.5°N, 53.3°W, 1,910 m), **d** proportion of Amazonian (vs. Orinoco) material within the terrigenous fraction of core GeoB3938-1 (12.3°N, 58.3°W, 1,972 m). The *black line* above the X-axis highlights the LIG period, as defined by benthic foraminiferal $\delta^{18}O$ values

austral summer insolation variations over the last 250 ka (Cheng et al. 2013). Therefore, the increase in austral summer insolation throughout the LIG (Fig. 2a) enhanced the ocean-land temperature gradient and moisture transport, thereby intensifying precipitation over the Amazon basin. Enhanced rainfall over South America and the adjacent ocean led to the reduced sea-surface salinities (SSS) that are observed close to the Amazon mouth and advected further northward at 115 ka BP compared to 125 ka BP (Fig. 1c). Strong stratification of surface waters indicated by planktic foraminiferal $\delta^{18}O$ records at 5°N and 9°N (not shown) supports this result.

While model and proxy data suggest increasing Amazon precipitation and runoff across the LIG, the core at 12°N exhibits a decrease in the relative proportion of Amazon (vs. Orinoco) material that is centered around ∼119 ka BP (Fig. 2d). The model also simulates reduced precipitation and runoff in the Orinoco basin at 115 ka BP compared to 125 ka BP (Fig. 1a, b), which agrees with the decrease in boreal summer insolation (Fig. 2a) but seems to disagree with the relative increase in Orinoco input at 12°N (Fig. 2d). In order to reconcile these findings, we suggest that surface ocean currents, which redistribute freshwater input from the Amazon

River, exert a strong influence on the signal at 12°N. Today, the North Brazil Current (NBC) carries Amazon freshwater towards the Caribbean Sea. However, the NBC retroflection deflects up to 70 % of the Amazon plume eastward between July and December (Lentz 1995). In a way similar to cold episodes of the last 30 ka (Wilson et al. 2011), we propose that the NBC retroflection was seasonally intensified or prolonged in duration at the end of the LIG, deflecting oceanward the plume of Amazon freshwater. Such changes would explain the relative decrease in Amazon material recorded around 119 ka BP at 12°N (Fig. 2d), despite increasing Amazon River discharge. The enhanced NBC retroflection simulated by the climate model (Fig. 1d) at 115 ka BP compared to 125 ka BP supports this hypothesis (Wilson et al. 2011). Our experiments isolating the impact of river runoff show that the 16 % increase in Amazon river discharge contributed to the plume of decreased SSS in the tropical Atlantic (Fig. 1e). As a result of the fresher North Atlantic, the position of the North Atlantic Deep Water formation region is shifted southward (Fig. 1f), inducing a significant ($p < 0.05$) 4 % decrease in deep water formation. This change in North Atlantic salinity and circulation led to changes in sea-surface temperatures (not shown), which may impact Northern hemisphere atmospheric circulation and ice sheet inception at the end of the LIG.

In summary, our model-data comparison indicates substantial shifts in South American hydrologic cycle and upper tropical Atlantic salinities that may impact the THC and North Atlantic climate at the end of the LIG.

References

Bakker P, Stone EJ, Charbit S, Gröger M, Krebs-Kanzow U, Ritz SP, Varma V, Khon V, Lunt DJ, Mikolajewicz U, Prange M, Renssen H, Schneider B, Schulz M (2013) Last interglacial temperature evolution—a model inter-comparison. Clim Past 9(2):605–619. doi:10.5194/cp-9-605-2013

Cheng H, Sinha A, Cruz FW, Wang X, Edwards RL, d'Horta FM, Ribas CC, Vuille M, Stott LD, Auler AS (2013) Climate change patterns in Amazonia and biodiversity. Nat Commun 4:1411. doi:10.1038/ncomms2415

Govin A, Chiessi CM, Zabel M, Sawakuchi AO, Heslop D, Hörner T, Zhang Y, Mulitza S (2014) Terrigenous input off northern South America driven by changes in Amazonian climate and the North Brazil current retroflection during the last 250 ka. Clim Past 10:843–862. doi:10.5194/cp-10-843-2014

Guyot JL, Jouanneau JM, Soares L, Boaventura GR, Maillet N, Lagane C (2007) Clay mineral composition of river sediments in the Amazon Basin. Catena 71(2):340–356. doi:10.1016/j.catena.2007.02.002

Krebs U, Timmermann A (2007) Tropical air-sea interactions accelerate the recovery of the atlantic meridional overturning circulation after a major shutdown. J Clim 20(19):4940–4956. doi:10.1175/jcli4296.1

Lentz SJ (1995) Seasonal variations in the horizontal structure of the Amazon Plume inferred from historical hydrographic data. J Geophys Res Oceans 100(C2):2391–2400. doi:10.1029/94jc01847

Tzedakis PC, Channell JET, Hodell DA, Kleiven HF, Skinner LC (2012) Determining the natural length of the current interglacial. Nat Geosci 5(2):138–141. doi:10.1038/ngeo1358

Wilson KE, Maslin MA, Burns SJ (2011) Evidence for a prolonged retroflection of the North
    Brazil current during glacial stages. Palaeogeogr Palaeoclimatol Palaeoecol 301(1–4):86–96.
    doi:10.1016/j.palaeo.2011.01.003
Yeager SG, Shields CA, Large WG, Hack JJ (2006) The low-resolution CCSM3. J Clim 19
    (11):2545–2566. doi:10.1175/jcli3744.1

Evaluation of Eemian and Holocene Climate Trends: Combining Marine Archives with Climate Modelling

Gerrit Lohmann, Ralph Schneider, Johann H. Jungclaus,
Guillaume Leduc, Nils Fischer, Madlene Pfeiffer and Thomas Laepple

Abstract In an attempt to assess trends of Holocene sea-surface temperature (SST), two proxies have been compiled and analyzed in light of model simulations. The data reveal contrasting SST trends, depending upon the proxy used to derive Holocene SST history. To reconcile these mismatches between proxies in the estimated Holocene SST trends, it has been proposed that the Holocene evolution of orbitally-driven seasonality of the incoming radiation is the first-order driving mechanism of the observed SST trends. Such hypothesis has been further tested in numerical models of the Earth system with important implications for SST signals ultimately recorded by marine sediment cores. The analysis of model results and alkenone proxy data for the Holocene indicate a similar pattern in temperature change, but the simulated SST trends underestimate the proxy-based SST trends by a factor of two to five. SST trends based on Mg/Ca show no correspondence with model results. We explore whether the consideration of different growing seasons and depth habitats of the planktonic organisms used for temperature reconstruction could lead to a better agreement of model results with alkenone data on a regional scale. We found that invoking shifts in the living season and habitat depth can remove some of the model–data discrepancies in SST trends. Our results indicate that modeled and reconstructed temperature trends are to a large degree only qualitatively comparable, thus providing at present a challenge for the interpretation of proxy data as well as the model sensitivity to orbital forcing.

G. Lohmann (✉) · M. Pfeiffer · T. Laepple
Alfred Wegener Institute, Helmholtz Centre for Polar and Marine Research,
Bremerhaven, Germany
e-mail: gerrit.lohmann@awi.de

G. Lohmann
Institute of Environmental Physics, University of Bremen, Bremen, Germany

R. Schneider · G. Leduc
Department of Geosciences, Institute of Geosciences, University of Kiel, Kiel, Germany

J.H. Jungclaus · N. Fischer
Max Planck Institute for Meteorology, Hamburg, Germany

© The Author(s) 2015
M. Schulz and A. Paul (eds.), *Integrated Analysis of Interglacial Climate Dynamics (INTERDYNAMIC)*, SpringerBriefs in Earth System Sciences,
DOI 10.1007/978-3-319-00693-2_6

Keywords Temperature trends · Holocene · Proxy data · Earth system models · Climate variability · Data-model comparison

1 Introduction

In order to examine sea-surface temperature (SST) trends caused by growing emissions of greenhouse gases and how they induce a significant impact on the Earth's climate, we need the knowledge about the variability of the natural system. Unfortunately, the instrumental record with a large-scale coverage of data goes back only to the time when human industrialisation started. Information beyond the instrumental record covering the last 150 years can only be obtained indirectly from two strategies. On the one hand, they can be derived from proxies that record past climate and environmental conditions. On the other hand, the past climate can be simulated using comprehensive models of the climate system under appropriate external forcing. Numerical climate models are clearly able to simulate a broad suite of phenomena in the current climate system, but their reliability on longer time-scales requires additional evaluation. Only climate records derived from paleoenvironmental proxies enable the test of these models because they provide records of climate variations that have actually occurred in the past.

2 Materials and Methods

a. Data: The proxy data work undertaken was focused on updating the global database for proxy-derived Holocene SST records, i.e., an SST synthesis based on alkenone-derived SST estimation, and a synthesis effort compiling the SST records derived from foraminiferal Mg/Ca was carried out (Leduc et al. 2010a).
b. Complex models: We use and evaluate a suite of atmosphere-ocean circulation models to evaluate the temperature evolution. We concentrate on ECHO-G (Lorenz and Lohmann 2004) and COSMOS (Fischer and Jungclaus 2010, 2011; Pfeiffer and Lohmann 2013), but use also data from time slice experiments as compiled in PMIP2 and PMIP3 (Lohmann et al. 2013).
c. Conceptual model and theoretical framework: A concept for the physical understanding of insolation-driven temperature variability on orbital timescales is developed. Based at this concept, the temperature evolution of the interglacials related to local insolation forcing is estimated (Laepple and Lohmann 2009).
d. Data-model and intermodel comparisons: We used several statistical techniques to compare data with models. We also participated on model intercomparisons (e.g., Lunt et al. 2013).

3 Key Findings

For the mid-Holocene, high obliquity results in more high-latitude summer insolation at the expense of low-latitude summer insolation. Obliquity explains most of
the variance in the annual insolation, and the effect is symmetric between the
hemispheres but asymmetric between the tropics and high latitudes. The seasonal
template model (Laepple and Lohmann 2009) largely reproduces the Holocene
temperature trends as simulated by coupled climate models and provides a theoretical framework for our project.

Our multi-proxy mapping reveals contrasting Holocene SST trends, depending
upon the proxy used. To reconcile these mismatches between proxies, we find that
foraminiferal Mg/Ca and alkenones paleothermometers may be skewed toward
specific seasons (Leduc et al. 2010a) (Fig. 1). Following on a seasonality
hypothesis, a first attempt to test and quantify the degree to which SST databases
are seasonally-skewed was conducted in data-model comparisons aiming at filtering
model output for different seasons and compare it to the SST database (Leduc et al.
2010a; Lohmann et al. 2013).

Using the coupled atmosphere-ocean general circulation model COSMOS with
applied orbital forcing, we investigate the climate evolution and variability of the last
two interglacial periods, the mid-Holocene (6 thousand years (ka) before present
(BP)) and the Last Interglacial (LIG) (125 ka BP). Earth's orbital parameters in these
two periods lead to an increase in the Northern Hemisphere's seasonal insolation

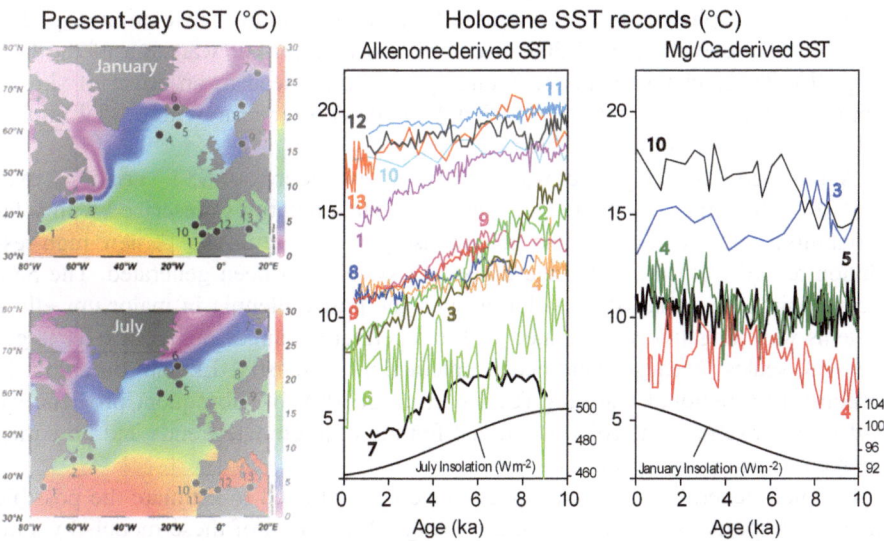

Fig. 1 *Left panels* modern-day seasonal anomalies in SST in the North Atlantic Ocean (in °C) and
locations of marine sediment cores corresponding to the records shown in *right panels*. *Right
panels* records for paleo-SST covering the last 10 ka and estimated from alkenone unsaturation
index and planktonic foraminifera Mg/Ca measurements. Adapted from Leduc et al. (2010a)

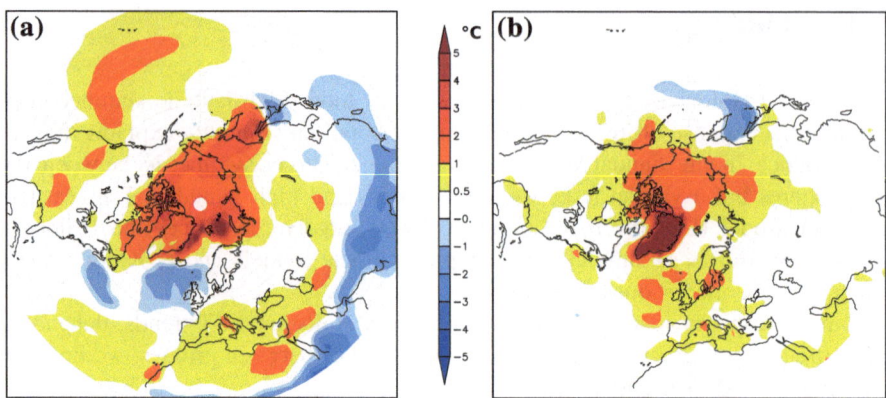

Fig. 2 Annual mean surface temperature anomalies for: **a** LIG minus the preindustrial control simulation, **b** LIG with reduced GIS by 1,300 m minus LIG. For the run with reduced GIS, the surface area and albedo have been consistently adapted

cycle. In high northern latitudes, an insolation-induced temperature increase is further enhanced by increased ocean heat transport and reduced sea-ice cover, resulting in an intensified ocean heat release (Fischer and Jungclaus 2010, 2011).

During the LIG, the northern high latitudes showed summer temperatures higher than those of the late Holocene, and a significantly reduced Greenland Ice Sheet (GIS). We perform sensitivity studies for the height and extent of the GIS at the beginning of the LIG (130 ka BP) using COSMOS. Our study shows that a strong Northern Hemisphere warming is caused by increased summer insolation (Fig. 2a). Reduced GIS elevation by 1,300 m contributes to a further warming of the LIG (Fig. 2b). These changed model boundary conditions lead to an effect of similar amplitude over high latitudes (up to 7 °C) compared to the run with orbital forcing only (Pfeiffer and Lohmann 2013).

Another goal of our project was to provide new records of SST variability over the last 3 millennia to evaluate Holocene transient simulations prescribed with solar variability, volcanic forcing, and greenhouse gas concentrations. A new high-resolution record from the Benguela upwelling system has been generated. The SST record pointed out that SST evolution over the past millennia in major upwelling systems were negatively correlated with temperature changes observed over larger regional scales because land-ocean interactions intensify Ekman pumping in response to a regional warming (Leduc et al. 2010b). In subsequent projects, we will further examine the role of external forcing onto climate trends and how they are recorded in proxy data.

Coupled general circulation models have been utilized to estimate the possible range of amplitudes for future climate change. Validation of these models by simulating interglacial climate states is essential for understanding the sensitivity of the climate system to external forcing. As a key issue in the 'Interdynamic' priority program of the German science foundation, we evaluated the interglacial dynamics by analyzing reconstructed and modeled SST trends *in tandem*. From our data-model comparison, we conclude that the SST sensitivity to orbital forcing seems to be underestimated in the models relative to the paleoclimate proxy data. More work is required to establish if such discrepancies can be caused by too simplistic interpretations of the proxy data, or by underestimated long-term feedbacks in climate models.

References

Fischer N, Jungclaus JH (2010) Effects of orbital forcing on atmosphere and ocean heat transports in Holocene and Eemian climate simulations with a comprehensive earth system model. Clim Past 6:155–168

Fischer N, Jungclaus JH (2011) Evolution of the seasonal temperature cycle in a transient Holocene simulation: orbital forcing and sea-ice. Clim Past 7:1139–1148

Laepple T, Lohmann G (2009) The seasonal cycle as template for climate variability on astronomical time scales. Paleoceanography 2:PA4201. doi:10.1029/2008PA001674

Leduc G, Schneider RR, Kim JH, Lohmann G (2010a) Holocene and Eemian sea surface temperature trends as revealed by alkenone and Mg/Ca paleothermometry. Quat Sci Rev 29:989–1004

Leduc G, Herbert C, Blanz T, Martinez P, Schneider R (2010b) Contrasting evolution of sea surface temperature in the Benguela upwelling system under natural and anthropogenic climate forcings. Geophys Res Lett 37:L20705

Lohmann G, Pfeiffer M, Laepple T, Leduc G, Kim JH (2013) A model-data comparison of the Holocene global sea surface temperature evolution. Clim Past 9(1–52):1807–1839. doi:10.5194/cp-9-1807-2013

Lorenz S, Lohmann G (2004) Acceleration technique for Milankovitch type forcing in a coupled atmosphere-ocean circulation model: method and application for the Holocene. Clim Dyn 23 (7–8):727–743

Lunt DJ, Abe-Ouchi A, Bakker P, Berger A, Braconnot P, Charbit S, Fischer N, Herold N, Jungclaus JH, Khon VC, Krebs-Kanzow U, Lohmann G, Otto-Bliesner B, Park W, Pfeiffer M, Prange M, Rachmayani R, Renssen H, Rosenbloom N, Schneider B, Stone EJ, Takahashi K, Wei W, Yin Q (2013) A multi-model assessment of last interglacial temperatures. Clim Past 9:699–717

Pfeiffer M, Lohmann G (2013) The last interglacial as simulated by an atmosphere-ocean general circulation model: sensitivity studies on the influence of the Greenland ice sheet. In: Lohmann G, Grosfeld K, Wolf-Gladrow D, Unnithan V, Notholt J, Wegner A (eds) Earth system science: bridging the gaps between disciplines perspectives from a multi-disciplinary Helmholtz Research School, Springer briefs in earth system sciences. Springer, Heidelberg, pp 57–64, doi:10.1007/978-3-642-32235-8

Holocene Environmental Variability in the Arctic Gateway

Robert F. Spielhagen, Juliane Müller, Axel Wagner, Kirstin Werner, Gerrit Lohmann, Matthias Prange and Rüdiger Stein

Abstract Environmental changes in the region connecting the Arctic Ocean and the northern North Atlantic were studied for the last 9,000 years (9 ka) by a combination of proxy-based paleoceanographic reconstructions as well as transient and time-slice simulations with climate models. Today, the area is perennially ice-covered in the west and ice-free in the east. Results show that sea-ice conditions were highly variable on short timescales in the last 9 ka. However, sea-ice proxies reveal an overall eastward movement of the sea-ice margin, in line with a decreasing influence of warm Atlantic Water advected to the Arctic Ocean. These cooling trends were rapidly reversed 100 years ago and replaced by the general warming in the Arctic. Model results show a consistently high freshwater input to the Arctic Ocean during the last 7 ka. The signal is robust against the Holocene cooling trend, however sensitive towards the warming trend of the last century. These results may play a role in the observed Arctic changes.

Keywords Arctic Ocean · Holocene · Sea ice · Atlantic Water · River run-off · Global warming

R.F. Spielhagen (✉)
Academy of Science and Literature, Mainz, Germany
e-mail: rspielhagen@geomar.de

R.F. Spielhagen · K. Werner
GEOMAR, Helmholtz Centre for Ocean Research Kiel, Kiel, Germany

J. Müller · A. Wagner · G. Lohmann · R. Stein
Alfred Wegener Institute, Helmholtz Centre for Polar and Marine Research, Bremerhaven, Germany

A. Wagner · M. Prange
University of Bremen, Bremen, Germany

© The Author(s) 2015
M. Schulz and A. Paul (eds.), *Integrated Analysis of Interglacial Climate Dynamics (INTERDYNAMIC)*, SpringerBriefs in Earth System Sciences, DOI 10.1007/978-3-319-00693-2_7

37

1 Introduction

The Arctic Ocean and sub-Arctic play a fundamental role in the global ocean-climate system. Siberian and North American rivers discharge enormous amounts of fresh-water to the Arctic Ocean which, in combination with cold atmospheric temperatures, enables the formation and persistence of the Arctic sea-ice cover which is severely threatened today by global warming. The Fram Strait between Greenland and Svalbard is the main "Arctic Gateway" for water mass exchange between the Arctic Ocean and the northern North Atlantic (Fig. 1). In the east, relatively warm (2–7 °C) and saline (S > 35) Atlantic Water flows northward. In the west, cold, sea-ice covered low-saline Arctic surface waters are exported to the Nordic Seas. Given the strong E-W contrasts, saline Atlantic Water is preconditioned for convective overturning and deepwater renewal. The position of the sea-ice margin is strongly controlled by the intensity and temperature of advected Atlantic Water but little is known about the Holocene (last ~12 ka) variability of these parameters. To fill this knowledge gap and analyze the amplitudes of natural Arctic climate variations on (sub-)millennial timescales, expressed as, e.g., temporal and spatial variations of Atlantic Water temperature and sea-ice cover, high-resolution sedimentary records from the Arctic Gateway were investigated by micropaleontological and geochemical methods. Simulations with a coupled atmosphere-ocean circulation model were used to investigate the role of river discharge and to extend the spatial range of sea-ice reconstructions which are originally derived from (local) sediment core data series.

2 Materials and Methods

Sediment investigations were performed on selected long cores with multidecadal resolution from the continental margin of the eastern Fram Strait (78.9°N–81.2°N; Fig. 2) and the shelf off East Greenland (73.1°N). The cores were subsampled to reach a (sub-)centennial resolution. The stratigraphy is based on a series of radio-carbon datings. For sedimentological, micropaleontological and isotopic investi-gations, sediment preparations and analyses followed standard procedures described by Werner et al. (2011, 2013). To investigate the organic-geochemical content, sediments were analyzed for total organic carbon, carbonate and biomarker composition according to procedures described by Müller et al. (2011, 2012).

Model simulations were performed with the coupled atmosphere-ocean general circulation model ECHO-G using acceleration techniques for the Holocene and the regional North Atlantic/Arctic Ocean–Sea Ice Model (NAOSIM). Details of the model set-up are given by Müller et al. (2011) and Wagner et al. (2011). Forcing occurred solely by solar variations from orbital parameters between 7 ka and 1800 CE and additionally by anthropogenic greenhouse gases thereafter. The high-res-olution NAOSIM simulations focus on the sea-ice distribution in the Greenland Sea and Fram Strait. Downscaling procedures were described by Müller et al. (2011) and Stärz et al. (2012).

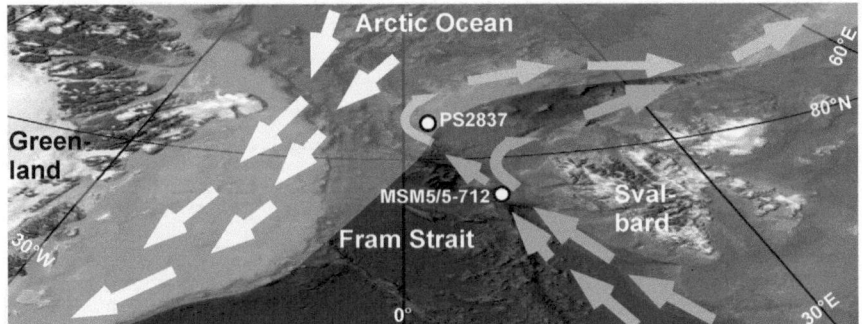

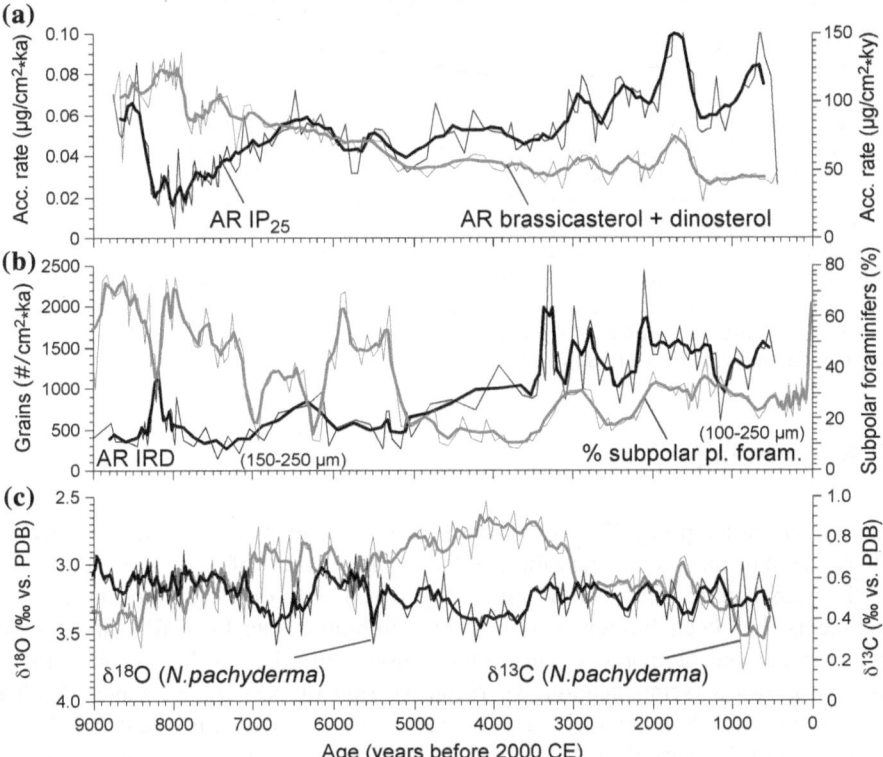

Fig. 1 Research area in the Fram Strait and proxy data sets from sediment core MSM5/5-712 for the last 9 ka. *Gray shading* denotes the modern average summer sea-ice cover. *White and gray arrows* in map indicate sea-ice drift and Atlantic Water advection, respectively. *Thick lines* in data series represent 3-point running means. **a** Accumulation rates (AR) of the sea-ice biomarker IP_{25} (*black*) vs. phytoplankton biomarkers (*gray*), **b** AR of terrigenous ice-rafted detritus (*black*) vs. relative abundance of subpolar planktic foraminifers (*gray*), **c** stable oxygen (*black*) and carbon isotopes (*gray*) of planktic foraminifers *Neogloboquadrina pachyderma* (125–250 µm fraction)

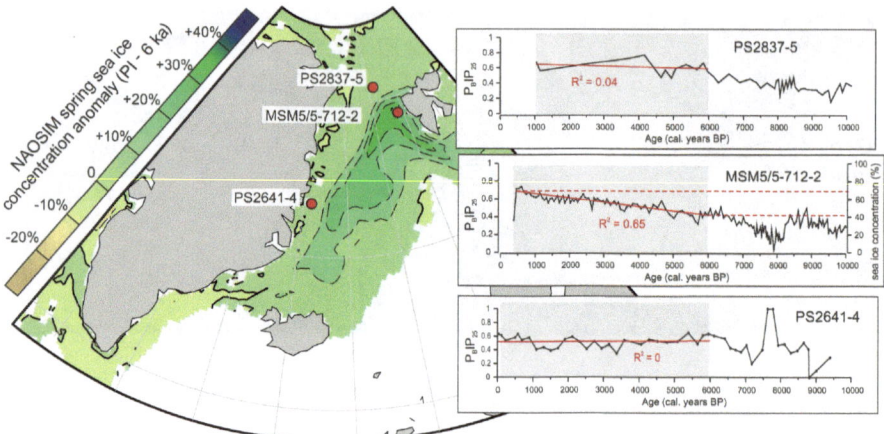

Fig. 2 NAOSIM-derived time-slice sea-ice anomaly for 6 ka to pre-industrial (PI) times in the Fram Strait area (map) and P_BIP_{25} (where B refers to the use of brassicasterol for calculating the PIP_{25} index) records of three sediment cores from the same area. *Gray shadings* and *red solid regression lines* highlight the individual P_BIP_{25} trends from 6 ka to PI times. Sea-ice concentration estimates for MSM5/5-712-2 are derived from the P_BIP_{25} vs. satellite sea-ice data correlation of Müller et al. (2011). Both proxy (P_BIP_{25}) and model results (*dark green* colours on the map) point to a 30 % sea-ice increase in the eastern Fram Strait between 6 ka and PI times (*red dashed lines* denote the increase in sea-ice concentrations from ca 40 % to 70 %). Only a minimum sea-ice increase is reconstructed (*solid red regression lines*)/modeled (*light green* colours on the map) for the northern and western Fram Strait core sites

3 Key Findings

The novel sea-ice proxy IP_{25}, a biomarker associated exclusively to diatoms living in sea ice, has opened new possibilities for reconstructions of past sea-ice coverage in the (sub-)Arctic. The method, originally based only on IP_{25} concentrations in sediments, has been further developed by combining data from IP_{25} and phytoplankton-derived biomarkers. This combination—resulting in the so-called PIP_{25} index—allows a (semi-)quantitative reconstruction of past sea-ice cover (Müller et al. 2009, 2011; Stein et al. 2012). This approach was first applied to core PS2837 from north of Svalbard (Fig. 1), close to the modern summer sea-ice margin (Müller et al. 2009). Variable contents of both biomarker types demonstrated that even the position of a seasonally fluctuating sea-ice margin can be reliably detected. A study on recent (surface) sediments in combination with numerical modeling results of the sea-ice distribution (Müller et al. 2011) showed that both methods successfully reproduce the degree and spatial distribution of the average ice coverage in the Fram Strait and Nordic Seas although in places local effects can distort the results.

Studies of biomarkers, microfossils, stable and radiogenic isotopes, as well as sedimentological proxies on core MSM5/5-712 (western Svalbard margin) allowed to reconstruct the variability of Atlantic Water advection (Fig. 1) to the Arctic and its effect on the regional ice coverage (Werner et al. 2011, 2013; Müller et al. 2012). IP_{25} accumulation rates are mostly low in late Early Holocene sediments and medium high at 7 to 3 ka, while phytoplankton biomarker contents decrease from 9 to 5 ka (Fig. 1). Proxy data and NAOSIM reconstructions (Fig. 2) suggest a general cooling trend and a successive southeastward shift of the sea-ice margin towards Svalbard since the Early Holocene (Müller et al. 2011, 2012), responding to the postglacial sea-level rise, a related onset of modern sea-ice production on the Siberian shelves (Werner et al. 2013), and to insolation-induced sea-ice and circulation changes. After 3 ka, biomarker accumulation rate peaks, highest contents of ice-rafted detritus, and a change in Nd isotope ratios of sediment leachates are evidence of a sea-ice margin rapidly advancing to and retreating from the core site (Müller et al. 2012; Werner et al. 2014). This interpretation is corroborated by microfossil and planktic isotope data. Generally high relative abundances of subpolar foraminifers until 5 ka indicate a strong influence of Atlantic Water as a near-surface water mass off western Svalbard, but low amounts thereafter demonstrate a drastic change to colder conditions (Werner et al. 2011, 2013). Low carbon and high oxygen isotope values of the polar planktic foraminifer species during the last 3 ka suggest a strengthened stratification due to a low-saline surface water layer. The drastic increase of subpolar species percentages in the last ~ 100 years indicates an equally strong increase of Atlantic Water temperatures and advection to the Arctic. Modern temperatures are unprecedented for the last 5 ka, probably as a response to global warming, and pose a severe threat to the Arctic sea-ice cover (Spielhagen et al. 2011).

Transient simulations with ECHO-G were analyzed to investigate the history of circum-Arctic river run-off and elucidate its possible role in the environmental changes observed in the Fram Strait (Wagner et al. 2011). The discharges, as calculated by the model, are driven by the difference between precipitation and evaporation and show a strong variability on annual to multi-centennial timescales. Results reveal that the discharge from Eurasian rivers increased slightly (2.1 ± 0.6 %) between 7 ka and 1800 CE while that of North American rivers decreased by 4.6 ± 0.6 %. In the Holocene, the total discharge remained at a constant level, while atmospheric temperatures decreased. This may have supported a stable halocline, as indicated by proxy data, and favored sea-ice formation. The last 100 years have seen a strong run-off increase from both continents (7.6 % for the total Arctic Ocean), in line with 20th century observations, however, accompanied by a rapid warming of the Arctic. These findings, together with the results from high-resolution sediment cores, clearly reveal extremely strong and rapid environmental changes during the Industrial Period in the Arctic which are unprecedented for several millennia before.

References

Müller J, Massé G, Stein R, Belt ST (2009) Variability of sea-ice conditions in the Fram Strait over the past 30,000 years. Nat Geosci 2(11):772–776

Müller J, Wagner A, Fahl K, Stein R, Prange M, Lohmann G (2011) Towards quantitative sea ice reconstructions in the northern North Atlantic: a combined biomarker and numerical modelling approach. Earth Planet Sci Lett 306(3–4):137–148

Müller J, Werner K, Stein R, Fahl K, Moros M, Jansen E (2012) Holocene cooling culminates in Neoglacial sea ice oscillations in fram strait. Quat Sci Rev 47:1–14

Spielhagen RF, Werner K, Sørensen SA, Zamelczyk K, Kandiano E, Budeus G, Husum K, Marchitto TM, Hald M (2011) Enhanced modern heat transfer to the Arctic by warm Atlantic Water. Science 331(6016):450–453

Stärz M, Gong X, Stein R, Darby DA, Kauker F, Lohmann G (2012) Glacial shortcut of Arctic sea-ice transport. Earth Planet Sci Lett 357–358:257–267

Stein R, Fahl K, Müller J (2012) Proxy reconstruction of Arctic Ocean sea ice history: from IRD to IP$_{25}$. Polarforschung 82:37–71

Wagner A, Lohmann G, Prange M (2011) Arctic river discharge trends since 7 ka BP. Global Planet Change 79:48–60

Werner K, Spielhagen RF, Bauch D, Hass HC, Kandiano E, Zamelczyk K (2011) Atlantic Water advection to the eastern Fram Strait - Multiproxy evidence for late Holocene variability. Palaeogeogr Palaeoclimatol Palaeoecol 308(3–4):264–276

Werner K, Spielhagen RF, Bauch D, Hass HC, Kandiano E (2013) Atlantic Water advection versus sea-ice advances in the eastern Fram Strait during the last 9 ka: Multiproxy evidence for a two-phase Holocene. Paleoceanography 28(2):283–295

Werner K, Frank M, Teschner C, Müller J, Spielhagen RF (2014) Neoglacial change in deep water exchange and increase of sea-ice transport through eastern Fram Strait: Evidence from radiogenic isotopes. Quat Sci Rev 92:190–207

Detecting Holocene Changes in the Atlantic Meridional Overturning Circulation: Integration of Proxy Data and Climate Simulations

Audrey Morley, David Heslop, Carsten Rühlemann, Stefan Mulitza, André Paul and Michael Schulz

Abstract The contribution of central water circulation to the Atlantic Meridional Overturning Circulation (AMOC) and its role in natural climate variability remain poorly understood. Limits in our knowledge are due to the scarcity of high resolution records from central water depth and the limited abilities of proxy parameters to reconstruct small changes in Holocene central water properties. We addressed these issues by combining paleoclimate modeling of 'fingerprints' (UVic ESCM version 2.8) that identify suitable locations and parameters to reconstruct past central water variability with the development and application of a new Mg/Ca-paleotemperature calibration for the benthic foraminifera *Hyalinea balthica*. The presented records demonstrate the important role of central water circulation in communicating regional climate signatures of various forcings (freshwater flux, solar variability, orbital parameters) onto a hemispheric or global scale via cross-gyre meridional heat transfer from high to low latitudes.

Keywords Paleoceanography · Ocean-atmosphere climate linkages · Atlantic meridional overturning circulation · Eastern North Atlantic central water · Solar variability · North Atlantic oscillation · Fingerprinting · Earth system climate model · Hosing experiment

A. Morley (✉)
School of Geography and Archaeology, National University of Ireland Galway,
Galway, Ireland
e-mail: audrey.morley@nuigalway.ie

D. Heslop
Research School of Earth Sciences, The Australian National University, Canberra, Australia

C. Rühlemann
Federal Institute for Geosciences and Natural Resources, Hannover, Germany

S. Mulitza · A. Paul · M. Schulz
MARUM—Center for Marine Environmental Sciences, University of Bremen,
Bremen, Germany

© The Author(s) 2015   43
M. Schulz and A. Paul (eds.), *Integrated Analysis of Interglacial Climate Dynamics (INTERDYNAMIC)*, SpringerBriefs in Earth System Sciences,
DOI 10.1007/978-3-319-00693-2_8

1 Introduction

The state of the North Atlantic Oscillation (NAO) in combination with subpolar gyre dynamics determines regional sea-surface heat loss and winter convection over the subpolar basin by modulating North Atlantic westerly wind stress and fresh water budgets (Hatun et al. 2005). During the winter, NAO modulated wind-stress favors the subduction of Subpolar Mode Water, which in turn comprises a large fraction of Eastern North Atlantic Central Water (ENACW) (Poole and Tomczak 1999). Flowing south along the eastern margin of the North Atlantic basin at densities between $\sigma_\theta = 27.3$ and 27.6 kg/m^3, ENACW circulation thus contributes to the ventilation of the thermocline and provides an 'oceanic tunnel', transmitting subpolar ocean-atmospheric climate signals to lower latitudes between 300 and 900 m water depth throughout the Holocene (Bamberg et al. 2010; Morley et al. 2011, 2014).

Little is known concerning the Holocene sensitivity, spatial and temporal response of central water circulation to abrupt or gradual climate forcing on multidecadal to millennial timescales. Major limitations on our ability to reconstruct central water properties arise from the scarcity of high resolution marine records from central water depth, a lack of central water proxies and limited investigation of central water circulation in numerical paleoclimate models. Using a combination of paleoclimate fingerprinting and proxy development we have developed a new method for reconstructing past central water properties on decadal to millennial time scales. This unified numerical and experimental approach has enabled us to evaluate the spatial and temporal response of Holocene central water circulation to various changes in climate forcing.

2 Material and Methods

For trace metal quantification, we analyzed up to 25 tests of the benthic foraminifera *Hyalinea balthica* from the 250 to 350 µm size fraction (mean: 20 individuals), using a modified reductive, oxidative cleaning protocol (Barker et al. 2003) and a Sector Field Inductively Coupled Plasma Mass Spectrometer (Thermo Element XR) at Rutgers Inorganic Analytical Laboratory. The long-term analytical precision of Mg/Ca ratios based on repeated analysis of three consistency standards was 1.6, 1.2 and 1.2 % RSD (relative standard deviation), respectively. The isotopic oxygen and carbon compositions ($\delta^{18}O$ and $\delta^{13}C$) of the foraminiferal shells were measured at the Stable Isotope Laboratory of the University of Bremen using a Finnigan MAT 251 mass spectrometer equipped with an automatic carbonate preparation device. Internal precision, based on replicates of a limestone standard, was better than ±0.07 % (VPDB).

Palaeoclimatic modeling was performed with the University of Victoria Earth System Climate Model (UVic ESCM version 2.8, Weaver et al. 2001) consisting of a two-dimensional atmospheric energy-moisture balance model, a dynamic-thermodynamic sea ice model and a three-dimensional ocean model (MOM2).

Specific forcing scenarios were developed to simulate internally and externally driven changes in Atlantic Meridional Overturning Circulation (AMOC) variability at a variety of time scales. Atlantic wide anomalous heat fluxes were imposed to create a basin scale temperature dipole that induced changes in northward heat transport (Heslop and Paul 2011). Additionally, idealized freshwater forcing scenarios were employed to reduce the strength of the simulated AMOC in a systematic manner. Spatial fingerprints of both internally and externally forced AMOC variability were estimated from the UVic ESCM runs using the metrics of Heslop and Paul (2012). In deriving these fingerprints attention was focused on the sensitivity, linearity and reversibility of seawater temperature responses to changing AMOC vigor.

3 Key Findings

Statistical fingerprints developed from the UVic ESCM model runs revealed the optimum conditions to detect and characterize short and long term changes in central water temperatures and their relationship to the AMOC (Heslop and Paul 2011, 2012). These fingerprints showed that central water temperature on the eastern boundary of the subtropical gyre should be a particularly sensitive and linear indicator of AMOC strength (Fig. 1).

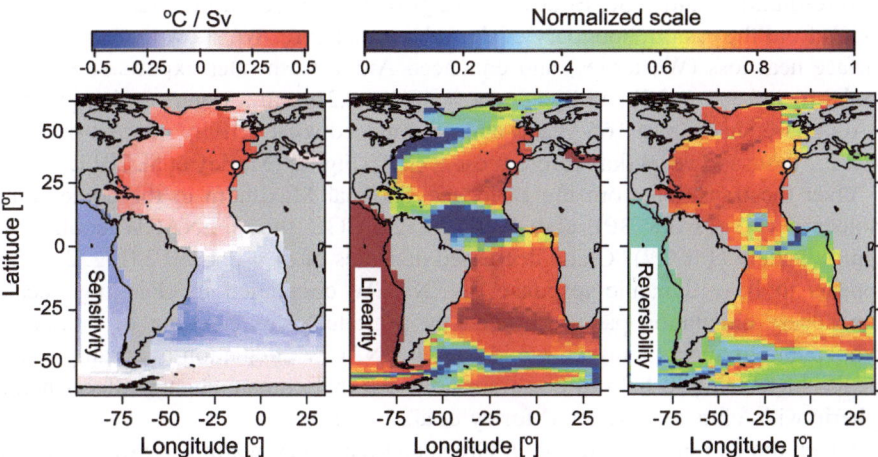

Fig. 1 Paleoclimate model fingerprints relating central water temperature properties at a depth of 860 m to AMOC strength. (*Left*) Sensitivities (units of °C/Sv) show the expected magnitude of the seawater temperature response to a given change in AMOC strength. *Red* (*blue*) shading corresponds to a positive (negative) relationship and *darker shading* indicates a larger amplitude response. (*Middle*) Linearity of the seawater temperature response to changing AMOC strength. *Red* (*blue*) shading on the normalized scale corresponds to a linear (nonlinear) relationship. (*Right*) Reversibility of seawater temperature change in response to changing AMOC strength. *Red* (*blue*) shading on the normalized scale indicates a reversible (nonreversible) relationship. Regions that show reversible behavior will not experience serious biases in water temperature when comparing periods of AMOC decline and recovery

The inclusion of freshwater forcing in the model runs to drive the AMOC state through a collection of partial hysteresis cycles, so-called first-order reversal curves (FORC), demonstrated that large areas of the Atlantic exhibit an asymmetrical temperature response to a declining and then recovering AMOC. At central water depths the temperature of the eastern subtropical gyre shows an approximately reversible relationship with AMOC evolution through a given FORC (Heslop and Paul 2012). This region is therefore suitable to derive high-fidelity records of past AMOC activity on the basis of proxy seawater temperature reconstructions. In contrast, other areas of the Atlantic exhibit strongly nonreversible responses, which imply that similar temperatures during the cooling and subsequent warming of a given episode cannot be assumed to correspond to the same AMOC strength.

We calibrated Mg/Ca ratios and $\delta^{18}O$ in tests of *H. balthica* to bottom water temperature (BWT) analyzing core tops from Indonesian and northeast Atlantic depth transects. The resulting calibration, Mg/Ca (mmol mol^{-1}) = (0.488 ± 0.03) BWT (°C), demonstrates *H. balthica* tests exhibit a temperature sensitivity four times higher than other studied deep-sea benthic foraminifera. Small calibration uncertainties [±0.7 °C, ±0.32 % VSMOW and ±0.69 psu (2σ)], allow estimation of seawaterdensity to <0.3 σ_θ units, which is sufficient to reconstruct changes in Holocene water mass properties and provenance (Rosenthal et al. 2011).

To estimate the contribution of central water circulation to the AMOC and its role in natural climate variability we applied the *H. balthica* calibration equation to high resolution multi- (GeoB6007-1) and gravity-cores (GeoB6007-2/OC437-7 24 GGC) from the eastern boundary of the subtropical gyre. We show that strong sea-surface heat loss (Westerlies) and enhanced Arctic freshwater exports resulted in cooler [(-0.8 ± 0.7) °C] and lighter ($\sigma_\theta = -0.3 \pm 0.2$) central waters during positive NAO phase shifts (past 165 years) and pronounced solar activity minima over the past 2.7 thousand years (ka) before present (BP) (Fig. 2) (Morley et al. 2011, 2014).

Over the transition from the Holocene Thermal Maximum to the cooler Late Holocene (3.3–2.6 ka BP), core OC437-7 24 GGC shows ENACW cooling of approximately (1 ± 0.7) °C and decreased densities of $\sigma_\theta = 0.4 \pm 0.2$ (Fig. 2). This appears to be a dynamic response of ENACW circulation to changing ocean-atmosphere circulation patterns forced by a gradual strengthening of latitudinal temperature gradients. Further, we identified ENACW circulation as an amplifying feedback, which will help to constrain how regional climate change affects hemispheric wide climate linkages (Morley et al. 2014).

Given the current acceleration in Greenland Ice Sheet melting, we evaluated the impact of early Holocene freshwater forcing on ENACW circulation. During times of enhanced Laurentide Ice Sheet melting (9.0–8.5 ka BP) central water production weakened. Additionally, two 150-year cooling events [(8.54 ± 0.2) ka BP and (8.24 ± 0.1) ka BP] provide evidence for an abrupt central water response to the drainage of Lake Agassiz (8.54 ka BP) in addition to the maximum AMOC slowdown at 8.2 ka BP (Fig. 2) (Bamberg et al. 2010). These observations provide a possible analogue for future effects on meridional heat transfer at central water depth from enhanced Greenland Ice Sheet melting and increases in Arctic freshwater export.

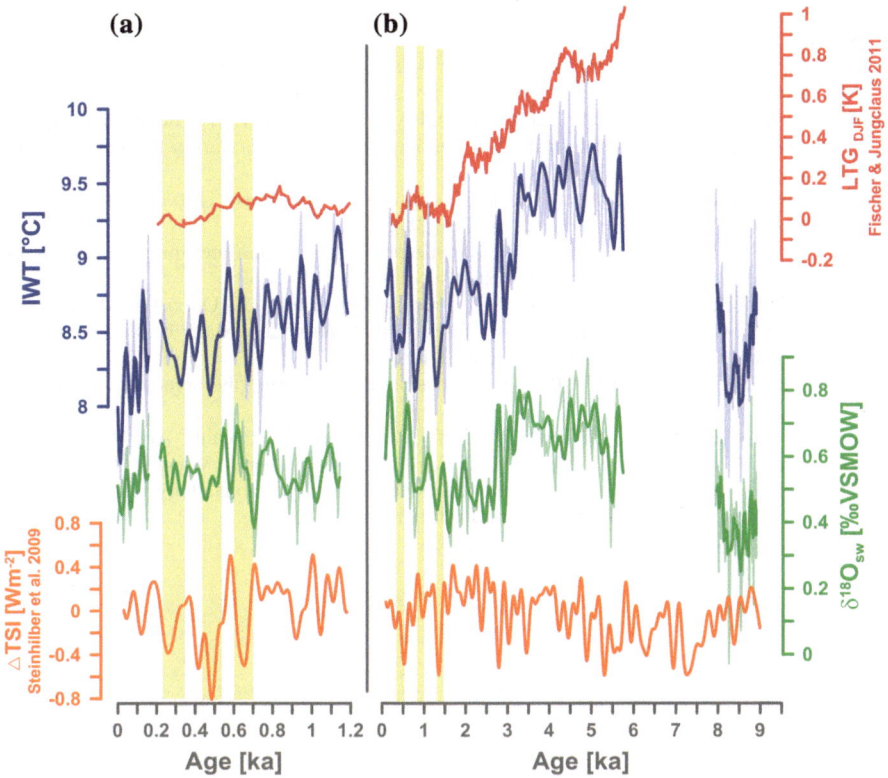

Fig. 2 Here we compare Mg/Ca based intermediate water temperatures (*IWT*) (*blue*) and calculated $\delta^{18}O_{sw}$ values derived from paired Mg/Ca–$\delta^{18}O_c$ measurements (*green*) with solar variability ΔTSI [Wm^{-2}] adapted from (Steinhilber et al. 2009) (*orange*) and reconstructed winter latitudinal temperature gradient (Dec–Jan–Feb) adapted from Fischer and Jungclaus (2011) (*red*). Bold lines refer to 60-year low-pass filters except for results based on OC437-7 24 GGC (0–6 ka BP) in panel b where *bold lines* show a 200-year low-pass filter. All *graphs* are plotted versus age. In panel a the three *yellow bars* indicate the timing of the Maunder (1), Spörer (2) and Wolf (3) solar minima (*left to right*) and in panel b the three *yellow bars* show the timing of the Maunder & Spörer (1), Wolf solar minima (2) and the Dark Ages Cold Period (3) (*left to right*).

References

Bamberg A, Rosenthal Y, Paul A, Heslop D, Mulitza S, Rühlemann C, Schulz M (2010) Reduced North Atlantic central water formation in response to early holocene ice-sheet melting. Geophys Res Lett 37:L1770

Barker S, Greaves M, Elderfield H (2003) A study of cleaning procedures used for foraminiferal Mg/Ca paleothermometry. Geochem Geophys Geosyst 4:8407

Fischer N, Jungclaus JH (2011) Evolution of the seasonal temperature cycle in a transient holocene simulation: orbital forcing and sea-ice. Clim Past 7:1139–1148

Hatun H, Sando AB, Drange H, Hansen B, Valdimarsson H (2005) Influence of the Atlantic subpolar gyre on the thermohaline circulation. Science 309:1841–1844

Heslop D, Paul A (2011) Can oceanic paleothermometers reconstruct the Atlantic multidecadal oscillation? Clim Past 7:151–159

Heslop D, Paul A (2012) Fingerprinting of the Atlantic meridional overturning circulation in climate models to aid in the design of proxy investigations. Clim Dynam 38:1047–1064

Morley A, Schulz M, Rosenthal Y, Mulitza S, Paul A, Rühlemann C (2011) Solar modulation of North Atlantic central water formation at multidecadal timescales during the late holocene. Earth Planet Sci Lett 308:161–171

Morley A, Rosenthal Y, DeMenocal P (2014) Ocean-atmosphere climate shift during the mid-to-late Holocene transition. Earth Planet Sci Lett 388:18–26

Poole R, Tomczak M (1999) Optimum multiparameter analysis of the water mass structure in the Atlantic Ocean thermocline. Deep Sea Res Part I 46:1895–1921

Rosenthal Y, Morley A, Barras C, Katz ME, Jorissen F, Reichart GJ, Oppo DW, Linsley BK (2011) Temperature calibration of Mg/Ca ratios in the intermediate water benthic foraminifer Hyalineabalthica. Geochem Geophys Geosyst 12:Q04003

Steinhilber F, Beer J, Fröhlich C (2009) Total solar irradiance during the holocene. Geophs Res Lett 36:L19704

Weaver AJ, Eby M, Wiebe EC, Bitz CM, Duffy PB, Ewen TL, Fanning AF, Holland MM, MacFadyen A, Matthews HD, Meissner KJ, Saenko O, Schmittner A, Wang H, Yoshimori M (2001) The UVic earth system climate model: model description, climatology, and applications to past, present and future climates. Atmos Ocean 39:361–428

Phase-Shift Between Surface Ocean Warming, Evaporation and Changes of Continental Ice Volume During Termination I Observed at Tropical Ocean Sediment Cores

Anton Eisenhauer, Christian Horn, Dirk Nürnberg, Thomas Blanz and Dieter Garbe-Schönberg

Abstract The hypothesis that the tropical oceans lead the global warming at the Termination I and II by $\sim 2,000$ to $\sim 3,000$ years (Visser et al. 2003) whereas melting of the northern continental ice masses is lacking behind challenges the Milankovitch theory of climate change and emphasizes the role of the tropics for global climate change. Although the simultaneous multi-proxy approach of planktonic foraminiferal Mg/Ca, $\delta^{18}O$ and $\delta^{44/40}Ca$ from tropical sediment core SO-164-03-4 ($16°$ $32.37'$ N; $72°$ $12.31'$ W; 2,744 m) from the Caribbean tend to confirm the observation by Visser et al. (2003) we interpret the shift between Mg/Ca and $\delta^{18}O$ in core SO-164-03-4 to be due to local changes in sea-surface salinity (SSS) variations triggered by glacial/interglacial related shifts of the Inter-tropical Convergence Zone (ITCZ).

Keywords Sea-surface temperature · Sea-surface salinity · Termination I · ITCZ · Foraminiferal geochemistry · Ca isotopes

1 Introduction

We investigated the important question in paleo-climatology to which degree the high northern latitudes or the tropics are triggering glacial/interglacial climate change. In more detail the proposed study is intended to examine the timing and phasing of ocean warming around Termination I verifying the hypothesis, put forward by Visser et al. in 2003. The inferences of Visser et al. (2003) are based on

A. Eisenhauer (✉) · C. Horn · D. Nürnberg
GEOMAR, Helmholtz Centre for Ocean Research Kiel, Kiel, Germany
e-mail: aeisenhauer@geomar.de

T. Blanz · D. Garbe-Schönberg
Department of Geosciences, Institute of Geosciences, University of Kiel, Kiel, Germany

© The Author(s) 2015
M. Schulz and A. Paul (eds.), *Integrated Analysis of Interglacial Climate Dynamics (INTERDYNAMIC)*, SpringerBriefs in Earth System Sciences,
DOI 10.1007/978-3-319-00693-2_9

49

sediment cores from the tropical West Pacific showing that the change in Mg/Ca measured in *G. ruber* leads its $\delta^{18}O_{foram}$ signal by $\sim 2,000$ to $\sim 3,000$ years (Fig. 1). However, $\delta^{18}O_{foram}$ and Mg/Ca values measured in foraminifera reflect sea-surface temperature (SST) but both are also controlled by their dependency on sea-surface salinity (SSS) which was quantified earlier by Nürnberg et al. (1996) and confirmed recently by Kisakürek et al. (2008). This implies that Mg/Ca reflect a local/regional signal whereas the $\delta^{18}O$ signal in planktonic foraminifera is interpreted by Visser et al. (2003) to reflect global rather than a local signal. Although SSS variations may also interfere with the $\delta^{18}O_{foram}$ signal the Visser et al. (2003) hypothesis can be tested by a multi-proxy approach simultaneously using $\delta^{18}O_{foram}$, Mg/Ca and $\delta^{44/40}Ca$ records applied to the same foraminiferal species of a sediment core from the Caribbean. Thereby, Mg/Ca and $\delta^{18}O$ is reflecting differences in both SST and SSS whereas $\delta^{44/40}Ca$ is supposed to be solely temperature-driven, independent from SSS variations (Gussone et al. 2004).

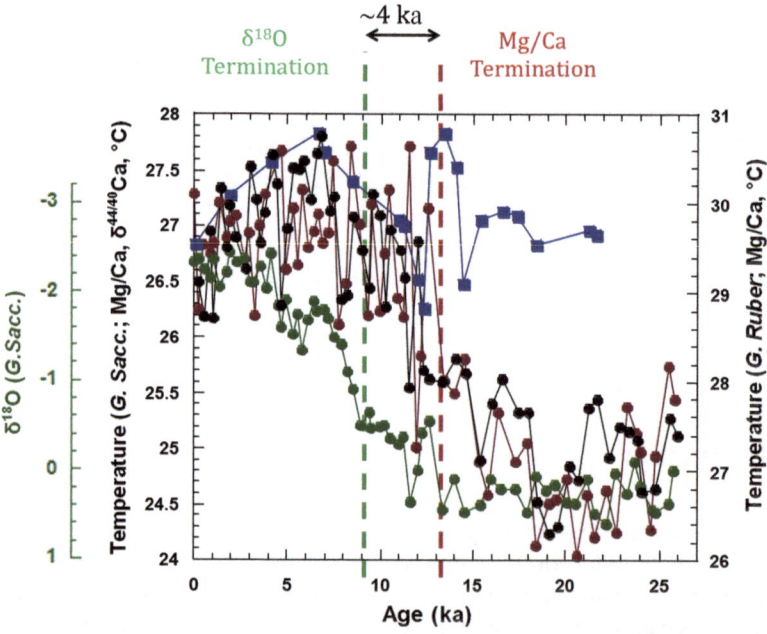

Fig. 1 This figure shows the Mg/Ca (*red G. sacc.*; *black G. ruber*), $\delta^{44/40}Ca$ (*blue G. sacc.*), and $\delta^{18}O$ records (*green G. sacc.*) values of two different foraminiferal species *G. ruber* and *G. sacculifer* of core SO164-03-4 as a function of time (ka = 1,000 years); *Y* Younger Dryas, *A* Allerød, *B* Bølling. Mg/Ca and $\delta^{44/40}Ca$ are in general accord after Termination I (*red dotted line* at ~ 13 ka BP) but differ to a larger extend during the Glacial. There are clear indications for a phase shift between Mg/Ca and $\delta^{18}O$ of ~ 4 ka between the Mg/Ca Termination and the lagging behind $\delta^{18}O$ Termination (*green dotted line* at ~ 9 ka BP) as it was reported earlier by Visser et al. (2003). Note, that the $\delta^{18}O$ values are roughly $\sim 2.5 \,$‰ too low when compared to the Mg/Ca record in the critical age range between 13 and 9 ka BP

2 Materials and Methods

Gravity core SO164-03-4 (16° 32.37′ N; 72° 12.31′ W; 2,744 m) was recovered during RV SONNE cruise SO164 in 2003 in the Columbian Basin of the central Caribbean Sea. All interpretations in this study are based on trace element and isotope measurements on the tropical spinose symbiont-bearing foraminifera species G. sacculifer and G. ruber. G. sacculifer mainly occurs during the summer season in water depths from 30 m down to ~50 m (Regenberg et al. 2009), whereas G. ruber tends to calcify in the warmer surface waters (shallower than 30 m when compared to G. sacculifer). The Mg/Ca-temperature calibrations of Nürnberg (2000) and of were applied to G. sacculifer and G. ruber, respectively.

Approximately 0.5–1.2 mg sample material usually consisting of 20–30 foraminiferal tests are needed for Mg/Ca analysis. The measurement of the Mg/Ca ratios followed standard protocols based on earlier studies (c.f. Kisakürek et al. 2008). The statistical uncertainty of the analytical procedure is in the order of 0.3 % representing one standard deviation of the mean. From the same sample aliquot about 15 single specimens of planktonic foraminifera are required for $\delta^{18}O$ measurements. The analytical reproducibility for the samples of this study is ~0.07 ‰ for $\delta^{18}O$.

Calcium isotope measurements are following the procedure of Heuser et al. (2002). The isotopic variations are expressed as $\delta^{44/40}Ca$-values using NIST SRM915a as standard material following the notation as proposed in Eisenhauer et al. (2004). In order to convert $\delta^{44/40}Ca$-values into temperature, the $\delta^{44/40}Ca$-temperature calibration of Hippler et al. (2006) was applied.

3 Key Findings

The Mg/Ca records of two different planktonic species although different in their absolute temperature estimates (Fig. 1) show that their glacial/interglacial Terminations I is at ~13 thousand years (ka) before present (BP) and at ~9 ka BP, respectively, for the $\delta^{18}O$-Termination I of G. sacculifer. Hence, there is general accord of a ~4 ka-offset from our observation and the earlier observation of Visser et al. (2003).

There is a general agreement of the $\delta^{44/40}Ca$- with the Mg/Ca-based temperature reconstructions (Fig. 1) in phase and time for the Holocene after Termination I. Notably, there is no systematic temperature difference between the Holocene and the Last Glacial in the $\delta^{44/40}Ca$-temperature record, hence no Termination can be attributed to this record. For the time interval before Termination I the $\delta^{44/40}Ca$-temperature relationship indicate no lower glacial temperatures and tend to be 1–2 °C warmer when compared to the corresponding Mg/Ca record. Latter observation is challenging pending future reconciliation.

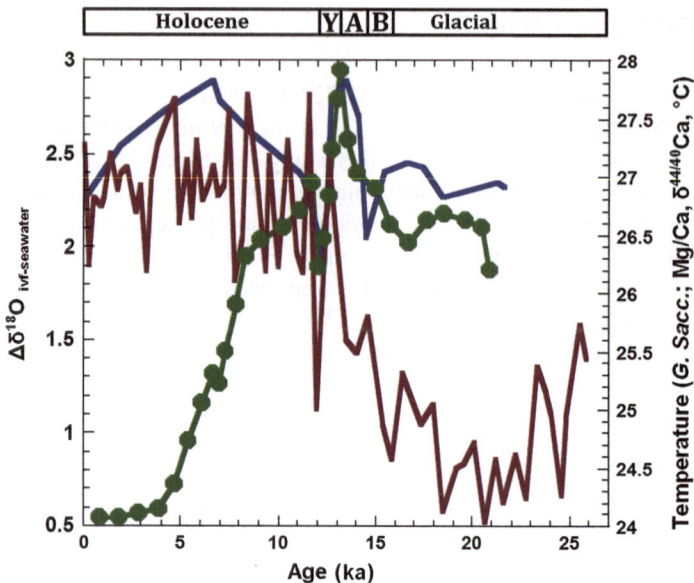

Fig. 2 Both Mg/Ca (*red*) and δ$^{44/40}$Ca (*blue*) measured on *G. sacculifer* of core SO164-03-4 are displayed as a function of time before present. It can be seen that there is no phase shift between the two records. Furthermore, it can be seen that both the δ$^{44/40}$Ca record as well as the Mg/Ca record show about the same pattern for the Holocene but are distinctively different for the Glacial. *Y* Younger Dryas, *A* Allerød, *B* Bølling. The measured δ^{18}O have been corrected for the glacial/interglacial ice volume change applying the δ^{18}O-record of (*ivf* ice volume free) and is reported as Δδ^{18}O$_{ivf-seawater}$ (*green curve*). It can be seen that highest relative salinities expressed as Δδ^{18}O$_{ivf-seawater}$ are predicted for the glacial and that maximum salinities are seen for the Allerød (A) and Younger Dryas (Y) period

In order to quantitatively account for the SSS change around Termination I we calculated the "ice volume free" (ivf) δ^{18}O composition of seawater; Δδ^{18}O$_{ivf-seawater}$ (Fig. 2) based on δ$^{44/40}$Ca-temperatures and normalized the values to the Holocene average. The Δδ^{18}O$_{ivf-seawater}$-values show that high Δδ^{18}O$_{ivf-seawater}$-values occurred until the Younger Dryas event and subsequently declined in the critical time interval between 13 and 9 ka BP towards the present day values (Fig. 2). In general, the δ^{18}O values in the time range between 10 and 15 ka tend to be up to ∼2.5 ‰ too heavy in order to match the Mg/Ca record.

Our initial working hypothesis was that SSS changes are a function of the local evaporation to precipitation ratio. Hence, any simultaneous local increase of SSS and SST is lowering the δ^{18}O-gradient to an extent lower than expected by a pure SST change alone. In the case of core SO164-03-4 the Δδ^{18}O$_{ivf-seawater}$-variations of up to 2.5 ‰ may reflect such a simultaneous SSS and SST change. Beside a glacial/ interglacial driven SST change, local SSS variations may be generated by any movement of the Inter-tropical Convergence Zone (ITCZ) center position either to the north or south, respectively, of its original position. In particular sediment core SO164-03-4 as well as the Visser et al. (2003) cores are from tropical regions and subject to the influence of the ITZC. Hence, it may be argued that the ITCZ position during the last glacial was shifted to the south but moved northward again at the glacial/interglacial transition to its present day position. In this regard the local salinity at the SO164-03-4 position increased gradually thereby shifting the $\delta^{18}O$ values to more positive values bringing the $\delta^{18}O$-record out of phase when compared to the other proxy records. Note, we consider our inferences still preliminary pending until quantification of the ITCZ movement over glacial/interglacial transitions.

References

Eisenhauer A, Nägler TF, Stille P, Kramers J, Gussone N, Bock B, Fietzke J, Hippler D, Schmitt AD (2004) Proposal for international agreement on ca notation resulting from discussion at workshops on stable isotope measurement held in Davos (Goldschmidt 2002) and Nice (EGS-AGU-EUG 2003). Geostand Geoanal Res 28(1):149–151

Gussone N, Eisenhauer A, Tiedemann R, Haug GH, Heuser A, Bock B, Nägler TF, Müller A (2004) Reconstruction of Caribbean Sea surface temperature and salinity fluctuations in response to the Pliocene closure of the Central American gateway and radiative forcing, using δ44/40Ca, $\delta^{18}O$ and Mg/Ca ratios. Earth Planet Sci Lett 227(3–4):201–214. doi:10.1016/j.epsl. 2004.09.004

Hippler D, Eisenhauer A, Nägler TF (2006) Tropical atlantic SST history inferred from Ca isotope thermometry over the last 140 ka. Geochim Cosmochim Acta 70:90–100

Heuser A, Eisenhauer A, Gussone N, Bock B, Hansen BT, Nägler TF (2002) Measurement of calcium isotopes ($\delta^{44}Ca$) using a multicollector TIMS technique. J Mass Spectrom 220:385–397

Kısakürek B, Eisenhauer A, Böhm F, Garbe-Schönberg D, Erez J (2008) Controls on shell Mg/Ca and Sr/Ca in cultured planktonic foraminiferan, *Globigerinoides ruber* (white). Earth Planet Sci Lett 273:260–269

Nürnberg D, Bijma J, Hemleben C (1996) Erratum assessing the reliability of magnesium in foraminiferal calcite as a proxy for water mass temperatures. Geochim Cosmochim Acta 60 (13):2483–2484

Nürnberg D (2000) Taking the temperature of past ocean surfaces. Science 289:1698–1699

Regenberg M, Steph S, Nürnberg D, Tiedemann R, Garbe-Schönber D (2009) Calibrating Mg/Ca ratios of multiple planktonic foraminiferal species with $\delta^{18}O$-calcification temperatures: paleothermometry for the upper water column. Earth Planet Sci Lett 278:324–336

Visser K, Thunell R, Stott L (2003) Magnitude and timing of temperature change in the Indo-pacific warm pool during deglaciation. Nat 421:152–155

Loop Current Variability—Its Relation to Meridional Overturning Circulation and the Impact of Mississippi Discharge

Dirk Nürnberg, André Bahr, Tanja Mildner and Carsten Eden

Abstract The dynamics of the Loop Current (LC) in the Gulf of Mexico (GoM) during transient climates and interglacials, and its interaction with changes in sea level, atmospheric circulation, and Mississippi River (MR) discharge were studied. Geochemical proxy records and numerical modeling indicate that LC eddy shedding and its related heat transport into the GoM increased during the deglaciation. The model simulations imply decreased LC eddy shedding at lowered sea levels, while transports through Yucatan and Florida straits increased due to the southward migration of the Intertropical Convergence Zone (ITCZ) and increased wind-driven transport in the North Atlantic. Consistent with the model, (isotope) geochemical proxy records from the northern GoM show glacial/interglacial amplitudes significantly larger than in the Caribbean and extreme cooling during the Last Glacial Maximum (LGM) due to the vanishing LC eddy shedding. Prominent deglacial melt water releases observed south and west of the MR delta are neither present in the northeastern GoM, nor in sea-surface salinity-records in the subtropical North Atlantic. The freshwater signals were either a regionally restricted phenomenon or due to changes in the isotopic composition of the discharged water. Our results question the impact of MR megadischarges on the large-scale overturning circulation.

Keywords Loop current · Eddy shedding · Gulf of Mexico hydrography · Mississippi discharge · Interglacials · Florida straits · Yucatan channel · Caribbean

D. Nürnberg (✉)
GEOMAR, Helmholtz Centre for Ocean Research Kiel, Kiel, Germany
e-mail: dnuernberg@geomar.de

A. Bahr
Institute of Geosciences, Goethe-University Frankfurt, Frankfurt am Main, Germany

T. Mildner · C. Eden
Institute of Oceanography, KlimaCampus, University of Hamburg, Hamburg, Germany

© The Author(s) 2015
M. Schulz and A. Paul (eds.), *Integrated Analysis of Interglacial Climate Dynamics (INTERDYNAMIC)*, SpringerBriefs in Earth System Sciences, DOI 10.1007/978-3-319-00693-2_10

55

1 Introduction

The Loop Current (LC) is the most prominent surface circulation feature in the Gulf of Mexico (GoM) flowing from the Caribbean through Yucatan Channel and Florida straits into the North Atlantic. Hence, the LC is a component of both, the western boundary current system of the North Atlantic and the basin- to global-scale meridional overturning circulation. It is therefore key to both present and past changes in the wind-driven basin-scale subtropical gyre and the global thermohaline circulation (e.g., Bryden et al. 2009; Lippold et al. 2012).

Special to the LC is the aperiodic shedding of anticyclonic meso-scale eddies, on time scales of 3–18 months (cf. Fig. 1a–c). The dynamics of the LC, in particular the mechanism of eddy separation and its possible link to external forcing (e.g. sea level and wind changes) is still not clear yet, but its role for the large-scale circulation in the North Atlantic is likely of importance (Mildner 2013). The focus of this study is therefore to understand the behavior of the LC and its eddy shedding during the past, in particular during transitional climate stages at the beginning of interglacials characterized by different sea level, atmospheric circulation, and wind stress forcing.

Together with the LC, the Mississippi River (MR) discharge influences the GoM surface hydrography. Flower et al. (2004) described distinct negative excursions in planktonic foraminiferal stable oxygen isotopes ($\delta^{18}O$) during Heinrich Event 1 and the Bølling/Allerød, which suggest Laurentide melt water discharge events into the GoM (e.g., Flower et al. 2004), although their large-scale impact on the Atlantic meridional overturning circulation (AMOC) remains ambiguous. Using paleoproxy records we study how these flooding events distribute in the GoM and adjacent ocean areas.

2 Materials and Methods

We discuss proxy data from the northern GoM (cores MD02-2575, MD02-2576, M78-181-3), Florida Straits (core KNR166-2-26), and Blake Outer Ridge (ODP Site 1058C) (Fig. 2). Details on the chronostratigraphies are given here and in Nürnberg et al. (2008), Kujau et al. (2010), Bahr et al. (2013), and Schmidt and Lynch-Stieglitz (2011).

Mg/Ca and $\delta^{18}O$ analyses were performed on the shallow-dwelling planktonic foraminifer *Globigerinoides ruber* white. Sea-surface temperature ($SST_{Mg/Ca}$) was determined using a common calibration. $\delta^{18}O_{seawater}$ was calculated combining foraminiferal $\delta^{18}O$ and $SST_{Mg/Ca}$ and subsequently, corrected for global ice volume variations yielding $\delta^{18}O_{ivf-sw}$ as an approximation for sea-surface salinity (SSS).

To investigate the terrigenous input into the northern GoM, we performed high resolution X-ray fluorescence (XRF) scanning converted to weight-% by using bulk

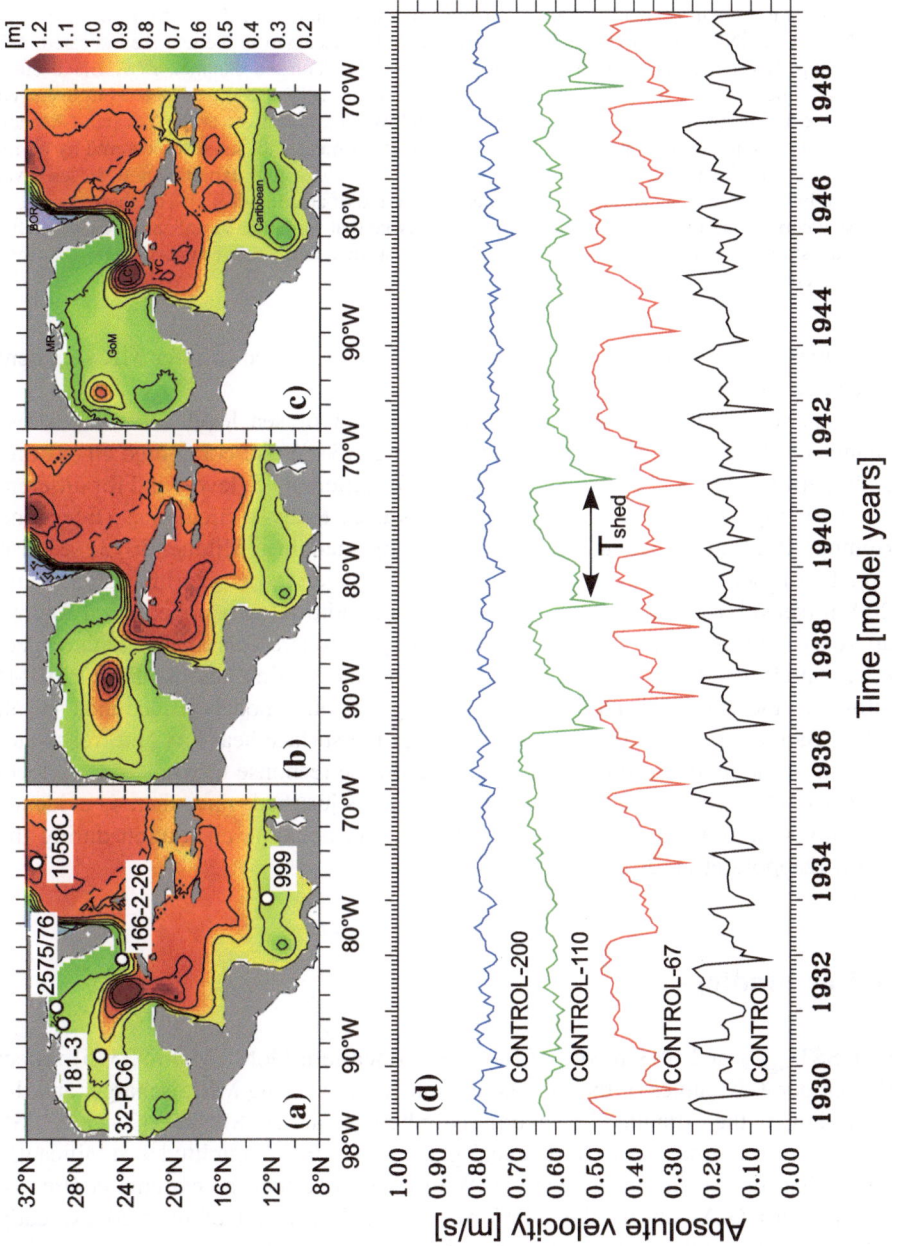

◄ **Fig. 1** **a–c** Eddy separation process in CONTROL shown by sea-surface elevation (in meters) depicted as composite from the last 20 model years (18 events in total); contour lines are 0.8 (1.2 m with 10 cm interval). **a** Scenario 3 months before shedding; **b** the eddy has just separated from the LC; **c** scenario 6 months after shedding. Locations for sediment cores MD02-2575/-76, M78-181-3, EN32-PC6, KNR166-2-26, ODP-1058C, and ODP-999 discussed in the text are indicated. *MR* Mississippi River, *LC* loop current, *FS* Florida straits, *YC* Yucatan channel, *BOR* Blake outer ridge. **d** Time series of spatially averaged current speed (in m/s) at 200 m water depth averaged across the region given by 22°N–24°N and 87°W–84°W, i.e., northwest of the Yucatan Channel. *Black line* present-day reference simulation. *Red line* experiment with sea level lowered by 67 m; *green line* experiment with sea level lowered by 110 m, *blue line* experiment with sea level lowered by 200 m. *Black arrow marks* the shedding period of an eddy (T_{shed}). The sharp decreases in speed correspond to eddy shedding from the LC. Note that for better visualization CONTROL-67 was shifted by 0.2 m/s, CONTROL-110 by 0.4 m/s and CONTROL-200 by 0.6 m/s, respectively

XRF measurements on selected samples. Potassium is used to infer MR sediment discharge (Kujau et al. 2010).

Numeric modeling concentrated on the impact of sea level and wind stress change on the LC and eddy shedding (Mildner 2013). We re-configured an existing eddy-permitting model of the North Atlantic for different sea levels and for different wind forcing. The model realistically reproduces today's circulation (Eden and Böning 2002), in particular the LC and its eddy shedding (Mildner 2013; see also Fig. 1a–c). Sea level was lowered by 67 m (Younger Dryas), 110 m (LGM), and 200 m (sensitivity experiment), respectively. In addition, we applied wind forcing appropriate to the LGM to our model by adding wind stress anomalies with respect to recent climate from 6 different coupled LGM model simulations of the PMIP-II effort to the standard recent wind forcing of our model (Braconnot et al. 2007; Barnier et al. 1995). We did not change the surface heat flux and freshwater forcing of the model in order to concentrate on the response to wind and sea level changes only. We believe this modeling strategy is justified, since there is evidence for only small changes in the North Atlantic thermohaline circulation during the LGM (Lippold et al. 2012).

3 Key Findings

Our $SST_{Mg/Ca}$ and SSS records from the northeastern GoM (MD02-2575) reflect the temporal dynamics of the LC, its relationship to varying MR discharge, and the evolution of the Atlantic Warmpool (Nürnberg et al. 2008). $SST_{Mg/Ca}$ and SSS records from the northern GoM reveal glacial/interglacial amplitudes significantly larger than in the Caribbean (Fig. 2). We hypothesize that the extreme cooling of the northern GoM during the LGM by ∼6 °C is a result of reduced LC eddy shedding and sluggish heat transport into the GoM. Considerable sea-surface freshening implies glacially enhanced river discharge.

The eddy-permitting model simulations support that LC eddy shedding was most likely absent during the LGM (Fig. 1). Lower sea level in the model (~ 110 and ~ 67 m) significantly reduces the number of eddies shed from the LC, as well as the oceanic heat transport into the GoM. With rising sea level, eddy shedding has gradually increased across the deglaciation and thereby, warmed the northern GoM up to its present state.

The simulated response to wind stress changes acts similar to the effect of sea level drop: A southward shift of the Intertropical Convergence Zone (ITCZ) and a strengthened atmospheric circulation during the LGM causes enhanced (wind-driven) Sverdrup transport within the Subtropical Gyre (Slowey and Curry 1995), leading to a strengthened Yucatan Channel and Florida Straits through flow. In response to the stronger transports, the eddy shedding decreases (Mildner 2013). Although PMIP models also simulate a stronger gyre circulation in the North Atlantic during the LGM, Lynch-Stieglitz et al. (2009) argue for a reduced Florida Straits transport during the LGM based on paleoceanographic proxy data pointing towards a contradiction, which needs further investigation.

A further effect of the changes in wind forcing in the model simulation is the southward expansion of the northern recirculation gyre in the North Atlantic Bight, and the southward shift of the zero line of wind stress curl. In response, both Gulf Stream and the Subtropical Gyre are shifted southwards (Mildner 2013). A highly dynamic Subtropical Gyre even during interglacial cool periods is suggested from ODP Site 1058C proxy data (Bahr et al. 2011, 2013) with a long-lasting subsurface warming during Marine Isotope Stage (MIS) 5 c-b, which most likely originated from intensified Ekman down welling in response to enhanced wind stress. The accumulation of warm and saline waters in the subsurface of the Subtropical Gyre might have contributed to the stabilization of northern hemisphere climate during MIS 5.

Across the deglaciation, the continuous increase in $SST_{Mg/Ca}$ in the northern GoM is accompanied by gradually decreasing fluvial sediment supply from the MR (Fig. 2), providing evidence that both the LC and the MR acted in concert. The deglacial SSS development, instead, is different: Prominent melt water signals ($\delta^{18}O_{ivf-sw}$) during Heinrich 1 and the Bølling/Allerød (Fig. 2) are observed only in cores south and west of the MR delta. In the northeastern GoM, instead, neither our SSS reconstruction nor the potassium record point to significant melt water discharge (Fig. 2). Apparently, the prominent freshwater signals were either a regionally restricted phenomenon or are due to changes in the isotopic composition of the discharge events. Since the $\delta^{18}O_{ivf-sw}$-record from Blake Outer Ridge is also devoid of such anomalies (Fig. 2), we question the importance of Laurentide Ice Sheet melt water routing via the MR for perturbing the AMOC.

During the Holocene, low potassium concentrations and positive $\delta^{18}O_{ivf-sw}$-values point to a negligible MR discharge when compared to LGM conditions. The rather gradual decline in $SST_{Mg/Ca}$ in the northeastern GoM after ~ 6 thousand

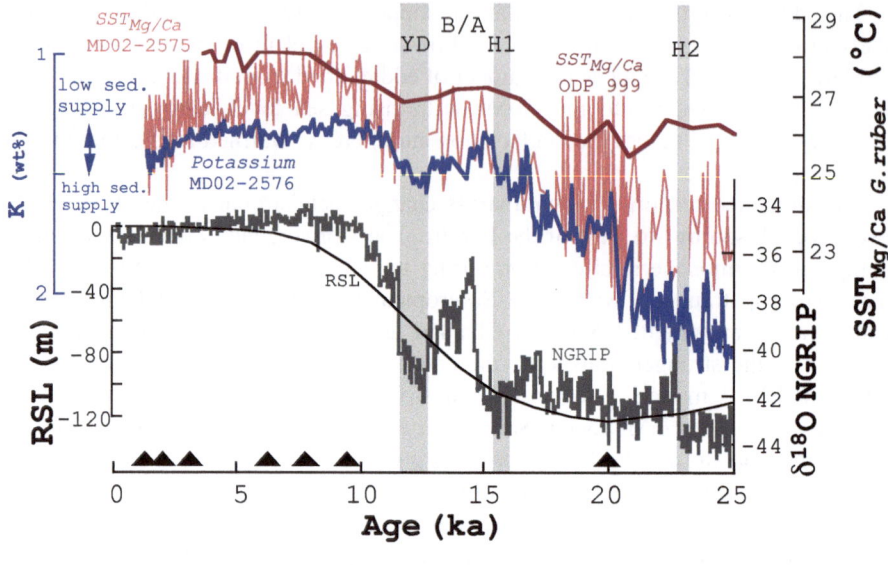

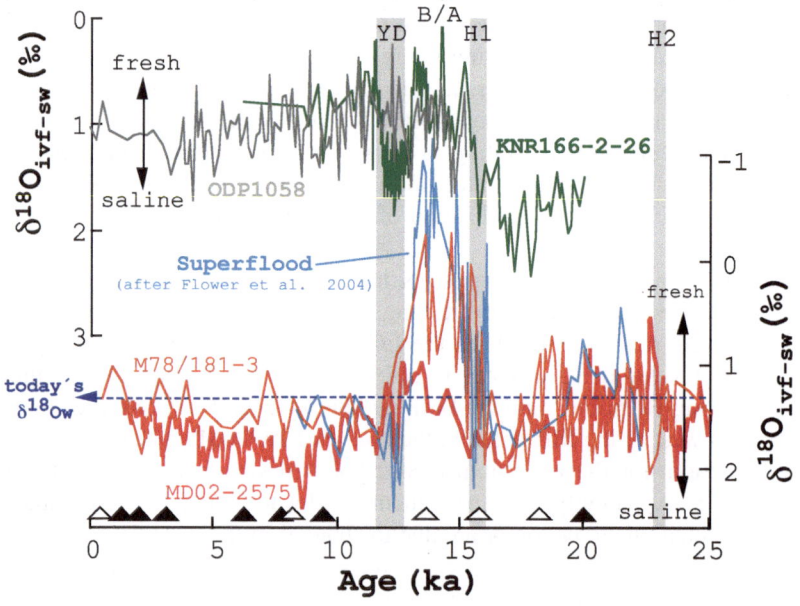

◄ **Fig. 2** Deglacial development of the LC and the MR discharge. *Left* GoM SST$_{Mg/Ca}$ record of IMAGES core MD02-2575 (*rosé*, 29°00.10′N, 87°07.13′W, 847 m depth) from DeSoto Canyon reflecting deglacial warming of the northeastern GoM in response to the northward migrating LC eddies (this study; Nürnberg et al. 2008; Ziegler et al. 2008) in comparison to (i) the Caribbean SST$_{Mg/Ca}$ record of ODP Site 999 (*brown*, 12°45′N, 78°44′W, 2827 m depth, all measured on the planktonic foraminifer *G. ruber* white; see Nürnberg et al. 2008), (ii) the potassium (K) record of IMAGES core MD02-2576 (*light blue*, 29°00.09′N, 87°07.14′W, 848 m) approximating terrigenous supply from the MR (Kujau et al. 2010), and (iii) the Waelbroeck et al. (2002) relative sea level curve (*black*). The stable oxygen isotope (δ^{18}O in ‰ VPDB) record of Greenland ice core NGRIP is used for stratigraphical control (*dark gray*). *Right* Various $\delta^{18}O_{ivf-sw}$ records approximating SSS changes from the northern GoM (MD02-2575, *red*, this study; M78-181-1, *orange*, 29°00.00′N, 88°20.00′W, 803 m depth, this study; EN32-PC6, *light blue*, 26°56.8′N, 91°20.0′W, 2280 m depth, after Flower et al. 2004), Florida Straits (KNR166-2-26, *green*, 24°19.60′N, 83°15.40′W, 547 m depth, Schmidt and Lynch-Stieglitz 2011), and *Blake Outer Ridge* (ODP Site 1058, *gray*, 31°41.00′N, 75°26.00′W, 2984 m, Bahr et al. 2013). The *blue* hatched line marks the today's $\delta^{18}O_{sw}$ value of the GoM (Lodico et al. 2006). *Shaded areas* indicate Heinrich Event 2 (H2), Heinrich Event 1 (H1), and the Younger Dryas (YD) cool event. *Black and open triangles* denote AMS^{14}C-datings at IMAGES core MD02-2575 (Nürnberg et al. 2008) and core M78-181-3 (this study), respectively

years (ka) before present (BP) (Fig. 2) is in line with an insolation-driven southward migration of the ITCZ, synchronous with the declining influence of the LC on the GoM hydrography. The decrease in SSS in the GOM is paralleled by wetter conditions over South Florida pointing to gradually increasing precipitation during the late Holocene.

Acknowledgements This work has been financed by the German Research Foundation (DFG) in the framework of the SPP1266 INTERDYNAMIK and IMAGES programs. Sample material was partly provided by the Integrated Ocean Drilling Program. The model integrations have been performed at University Kiel and Deutsches Klimarechenzentrum (DKRZ), Hamburg.

References

Bahr A, Nürnberg D, Schönfeld J, Garbe-Schönberg D (2011) Hydrological variability in Florida straits during Marine isotope stage 5 cold events. Paleoceanography 26(2):PA2214

Bahr A, Nürnberg D, Grützner J, Karas C (2013) Millennial-scale versus long-term dynamics in the surface and subsurface of the western North Atlantic subtropical gyre during Marine isotope stage 5. Global Planet Change 111:77–87. doi:10.1016/j.gloplacha.2013.08.013

Barnier B, Siefridt L, Marchesiello P (1995) Thermal forcing for a global ocean circulation model using a three-year climatology of ECMWF analyses. J Mar Syst 6(4):363–380. doi:10.1016/0924-7963(94)00034-9

Braconnot P, Otto-Bliesner B, Harrison S, Joussaume S, Peterschmitt JY, Abe-Ouchi A, Crucifix M, Driesschaert E, Fichefet T, Hewitt C, Kageyama M, Kitoh A, Laîné A, Loutre MF, Merkel U, Marti O, Ramstein G, Valdes P, Weber SL, Yu Y, Zhao Y (2007) Results of PMIP2 coupled simulations of the mid-holocene and last glacial maximum—part 1: experiments and large-scale features. Clim Past 3(2):261–277. doi:10.5194/cp-3-261-2007

Bryden HL, Mujahid A, Cunningham SA, Kanzow T (2009) Adjustment of the basin-scale circulation at 26 degrees N to variations in gulf stream, deep western boundary current and Ekman transports as observed by the rapid array. Ocean Sci 5(4):421–433

Eden C, Böning C (2002) Sources of eddy kinetic energy in the Labrador Sea. J Phys Oceanogr 32 (12):3346–3363. doi:10.1175/1520-0485

Flower BP, Hastings DW, Hill HW, Quinn TM (2004) Phasing of deglacial warming and laurentide ice sheet melt water in the gulf of Mexico. Geology 32(7):597–600

Kujau A, Nürnberg D, Zielhofer C, Bahr A, Röhl U (2010) Mississippi River discharge over the last ~560,000 years—indications from X-ray fluorescence core-scanning. Palaeogeogra Palaeocl 298:311–318. doi:10.1016/j.palaeo.2010.10.005

Lippold J, Luo Y, Francois R, Allen SE, Gherardi J, Pichat S, Hickey B, Schulz H (2012) Strength and geometry of the glacial Atlantic meridional overturning circulation. Nat Geosci 1–4. doi:10.1038/ngeo1608

Lodico JM, Flower BP, Quinn TM (2006) Subcentennial-scale climatic and hydrologic variability in the Gulf of Mexico during the early Holocene. Paleoceanography 21:PA3015. doi:10.1029/2005PA001243

Lynch-Stieglitz J, Curry WB, Lund DC (2009) Florida straits density structure and transport over the last 8,000 years. Paleoceanography 24(3):PA3209

Mildner TC (2013) Past and present ocean dynamics in the subtropical Atlantic. Dissertation, University of Hamburg. urn:nbn:de:gbv:18-64014

Nürnberg D, Ziegler M, Karas C, Tiedemann R, Schmidt MW (2008) Interacting loop current variability and Mississippi river discharge over the past 400 kyr. Earth Planet Sci Lett 272 (1–2):278–289

Schmidt MW, Lynch-Stieglitz J (2011) Florida straits deglacial temperature and salinity change: implications for tropical hydrologic cycle variability during the Younger Dryas. Paleoceanography 26:PA4205. doi:10.1029/2011PA002157

Slowey NC, Curry WB (1995) Glacial/interglacial differences in circulation and carbon cycling within the upper western North Atlantic. Paleoceanography 10(4). doi:10.1029/95PA01166

Waelbroeck C, Labeyrie L, Michel E, Duplessy JC, McManus JF, Lambeck K, Balbon E, Labracherie M (2002) Sea-level and deep water temperature changes derived from benthic foramifera isotopic records. Quat Sci Rev 21(1–3):295–305. doi:10.1016/S0277-3791(01)00101-9

Ziegler M, Nürnberg D, Karas C, Tiedemann R, Lourens LJ (2008) Persistent summer expansion of the Atlantic warm pool during glacial abrupt cold events. Nat Geosci 1:601–605. doi:10.1038/ngeo277

Hydroclimatic Variability in the Panama Bight Region During Termination 1 and the Holocene

Matthias Prange, Silke Steph, Huadong Liu, Lloyd D. Keigwin and Michael Schulz

Abstract A transect of sediment cores from high-sedimentation rate locations from the Panama Bight (eastern tropical Pacific) in combination with climate model experiments provides an opportunity to improve our understanding of the role of the tropical hydrologic cycle as a potential driver of global climate change during the Holocene and Termination 1. The reconstruction of regional sea-surface salinity patterns suggests the development of an anomalous precipitation dipole in the tropical eastern Pacific during Heinrich Stadial 1 (H1) with reduced rainfall over the western Panama Basin and off Costa Rica, and wetter conditions along the Colombian coast. Freshwater hosing experiments with the climate model CCSM3, mimicking the climatic reorganizations during H1, capture this precipitation dipole, while showing no change in the Atlantic-to-Pacific water vapor flux in response to a slowdown of the Atlantic meridional overturning circulation (AMOC). We conclude that the cross-isthmus vapor flux feedback on AMOC variations is negligible.

Keywords Termination 1 · Heinrich stadial 1 · Holocene · Moisture transport · Atlantic meridional overturning circulation · Panama bight · Sea-surface salinity · Global climate modeling

1 Introduction

Today, a significant portion of rainfall in the eastern tropical Pacific is attributed to the atmospheric transport of water vapor from tropical Atlantic/Caribbean sources via the northeasterly trade winds that cross Central America (e.g., Benway and Mix

M. Prange (✉) · S. Steph · H. Liu · M. Schulz
MARUM—Center for Marine Environmental Sciences and Faculty of Geosciences, University of Bremen, Bremen, Germany
e-mail: mprange@marum.de

L.D. Keigwin
Geology and Geophysics Department, Woods Hole Oceanographic Institution, Woods Hole, USA

© The Author(s) 2015
M. Schulz and A. Paul (eds.), *Integrated Analysis of Interglacial Climate Dynamics (INTERDYNAMIC)*, SpringerBriefs in Earth System Sciences, DOI 10.1007/978-3-319-00693-2_11

63

2004). This net export of freshwater helps to maintain relatively high salinities within the Atlantic and has been argued to exert a strong influence on the strength and stability of the Atlantic meridional overturning circulation (AMOC) (e.g., Zaucker et al. 1994; Romanova et al. 2004). So far, only three paleoceanographic studies were devoted to past changes in the cross-isthmus vapor transport during the Late Quaternary (Benway et al. 2006; Leduc et al. 2007; Pahnke et al. 2007). In these studies sea-surface salinity (SSS) reconstructions from the eastern tropical Pacific were used to infer changes in the Atlantic-to-Pacific moisture flux. Reconstructed local salinity increases (decreases) were basically interpreted as decreased (increased) cross-isthmus moisture transport, which however led to seemingly contradictory results, in particular during Heinrich Stadial 1 (H1). Increasing eastern Pacific SSS off Costa Rica (Benway et al. 2006; Leduc et al. 2007) and decreasing SSS in the Panama Bight (Pahnke et al. 2007) were later reconciled by postulating the development of an anomalous precipitation dipole in the eastern tropical Pacific during H1, similar to a modern La Niña situation (Prange et al. 2010). This analog would be associated with reduced Atlantic-to-Pacific vapor export (Schmittner et al. 2000) and hence a positive feedback on the H1 AMOC slowdown due to anomalous Atlantic Ocean freshening. In the framework of our study, a transect of sediment cores along the Colombian and Panamanian margins in combination with climate model studies provides an opportunity to improve our understanding of the role of the tropical hydrological cycle as a potential driving force for global climate change through Termination 1 and the Holocene.

2 Materials and Methods

In order to reconstruct past variations of the hydroclimate in the Panama Bight region, we use high-deposition rate sediment cores retrieved during R/V Knorr cruise 176-2 in 2004 (Fig. 1a; www.marine.whoi.edu/kn_synop.nsf). Located along the rim of the basin, these cores allow for the reconstruction of temporal changes in spatial hydroclimatic patterns by means of well-established geochemical methods. Coring sites from the Colombian margin are ideally located to sensitively monitor changes in continental river runoff. Using stable-isotope analyses on planktonic foraminifera (*Globigerina ruber*, *Neogloboquadrina dutertrei*) in combination with alkenone sea-surface temperatures (SSTs; Mg/Ca SST reconstructions were not possible due to the sparsity of *G. ruber* in most samples), a history of SSS changes and thermocline depth in the Panama Bight is developed for Termination 1 and the Holocene.

The proxy studies are accompanied by simulations of Holocene and deglacial (H1) climate states, using the atmosphere-ocean general circulation model CCSM3 (Collins et al. 2006). A reliable simulation of the regional climatic features in the eastern tropical Pacific with correct position of the Choco Jet—a low-level westerly wind jet centered at 5°N which transports Pacific moisture towards Colombia—requires the use of a relatively high spatial resolution of the atmospheric component. We therefore employed the T85 (1.4° transform grid) version of the model.

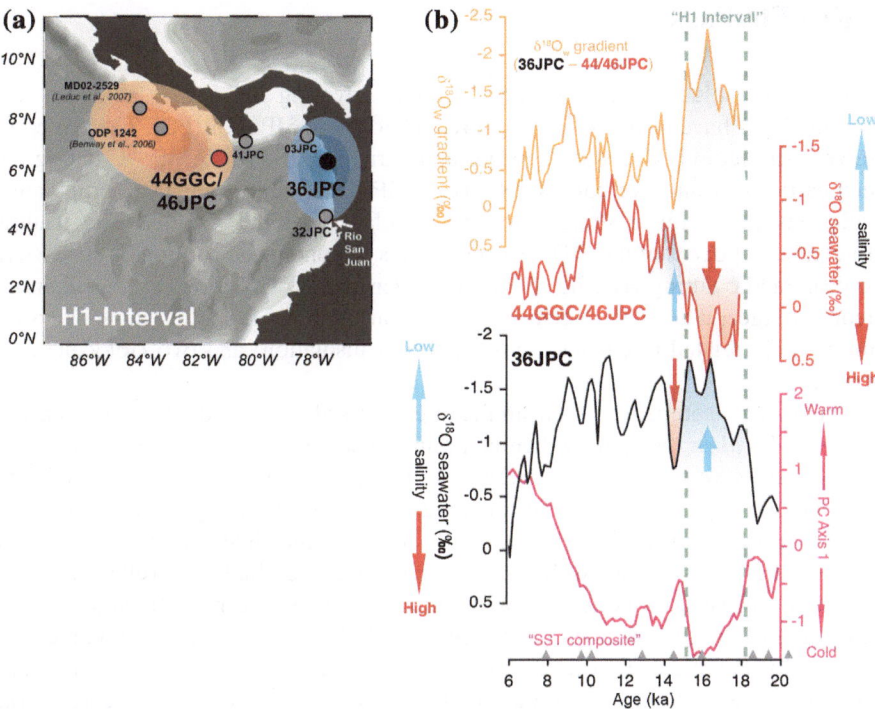

Fig. 1 a Locations of sediment cores collected during Knorr Cruise KNR176-2, which were analyzed. Cores used in previous studies (Benway et al. 2006; Leduc et al. 2007) are also shown. The location of the Rio San Juan delta is marked by an *arrow*. The H1 anomalous SSS or rainfall dipole is schematically illustrated by *blue* (wet) and *red* (dry) shading. **b** From *bottom* to *top*: leading principal component of 44GGC/46JPC, 36JPC, and 32JPC alkenone SST records; sea-level corrected $\delta^{18}O_{seawater}$ records of 36JPC and 44GGC/46JPC; difference between the two $\delta^{18}O_{seawater}$ records. *Grey triangles* mark AMS ^{14}C dates for core 32JPC, calibrated using Calib5.0.1-Marine04 without local offset

A modern control run shows that the model is able to successfully simulate the annual cycle of the regional wind pattern including strong northeast trades in late winter and early spring, and a large transport of water vapor from the Pacific to western Colombia with the Choco Jet during summer (not shown).

Freshwater hosing experiments under different background climatic conditions (pre-industrial, 8.5 thousand years (ka) before present (BP) early Holocene, and Last Glacial Maximum) were carried out to identify possible feedbacks between AMOC strength and tropical atmospheric vapor transports, with a particular eye towards the "8.2 ka BP event" (early Holocene experiment) and H1 (glacial experiment). In these hosing experiments, freshwater at a rate of 0.2 Sv is injected into the northern North Atlantic for 400 years leading to substantial weakening of the AMOC.

3 Key Findings

Three high-resolution alkenone SST records covering the entire Holocene and Termination 1 have been completed (44GGC/46JPC, 36JPC, 32JPC). These records are very similar in both absolute values and variability. So far, AMS ^{14}C dating has been carried out only for core KNR176-2 32JPC, but almost identical temperature evolutions allow for a straightforward correlation of the records. The leading principal component of the SST records reveals a pronounced cooling of the eastern tropical Pacific during H1 (Fig. 1b) in the order of 1–2 °C. Moreover, the alkenone records suggest a delayed deglacial warming in the region (starting at ∼11 ka BP and possibly related to changes in local winter insolation) and two Holocene cold events.

The oxygen isotope records from the three cores all reveal a deglacial decrease in $\delta^{18}O$ of *G. ruber*, but with different timing. The sea-level corrected $\delta^{18}O_{seawater}$ record (indicative of local SSS changes) from 32JPC, which is located close to the river mouth of Rio San Juan, indicates higher SSS during the glacial period and a freshening trend during H1 (not shown). At the same time, the $\delta^{18}O$ difference between the thermocline dweller *N. dutertrei* and the shallow dweller *G. ruber* increases (not shown), consistent with enhanced upper-ocean stratification due to freshening of the surface layer (cf. Steph et al. 2009). The results agree with earlier findings by Pahnke et al. (2007) based on δD in alkenones from the same core. Moreover, the comparison of the sea-level corrected $\delta^{18}O_{seawater}$ records from different sites indicates a pronounced "salinity dipole" in the tropical eastern Pacific during H1, with enhanced SSS (and weaker upper-ocean stratification) in the western Panama Basin (core 44GGC/46JPC) and reduced SSS (and stronger stratification) near the Colombian coast (core 36JPC; Fig. 1b). These results corroborate the precipitation dipole previously hypothesized for that region (Prange et al. 2010).

The CCSM3 model results help interpret the proxy records and set them into a large-scale dynamical context. Independent of the background climatic state, all hosing experiments capture a precipitation-anomaly dipole in the Panama Bight region in response to a weakening of the AMOC, with enhanced rainfall over western Colombia and reduced rainfall over the Gulf of Panama and west of Costa Rica (significant at the 0.05 level according to a t-test). The simulation of this hydroclimatic "fingerprint" lends confidence to the model. As an example, Fig. 2a shows the net precipitation response to a substantial AMOC weakening under preindustrial conditions.

The annual mean net moisture transports from the Atlantic to the Pacific were computed along the 6 °N–14 °N segment for all experiments based on daily model output. Compared to the pre-industrial simulation, the moisture transport at the Last Glacial Maximum decreases by 16 % (from 0.31 to 0.26 Sv). An increase of northeasterly trades is overcompensated by a lower atmospheric moisture content. The net vapor transport in the early Holocene run is very similar to the pre-industrial value (0.28 Sv). The net moisture transports across Central America for

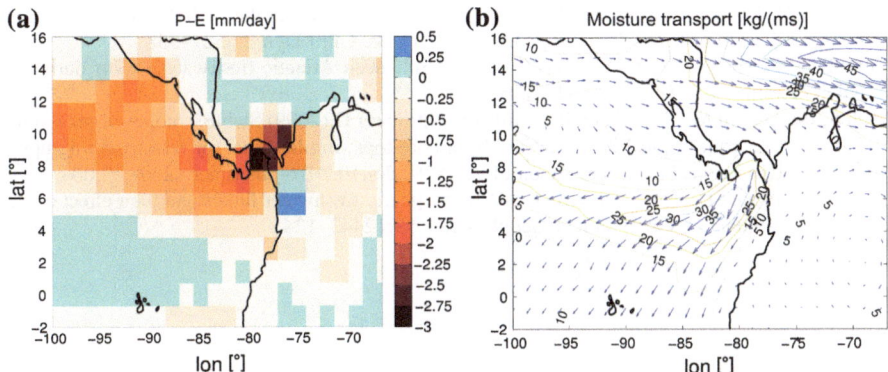

Fig. 2 a Annual mean P–E (precipitation minus evaporation; mm d^{-1}) response to a $\sim$50 % AMOC slowdown under preindustrial boundary conditions. **b** Corresponding changes in annual mean vertically integrated moisture transport (contour *lines* indicate the magnitude of transport; kg m^{-1} s^{-1})

the pre-industrial, early Holocene and glacial basic states show no differences (<0.01 Sv) compared to their freshwater hosing counterparts. As shown in Fig. 2b, slightly enhanced vapor fluxes across Panama are compensated by eastward flux anomalies further north. We therefore conclude that the cross-isthmus vapor flux feedback on AMOC variations is negligible. Moreover, our results suggest that changes in the rainfall and SSS patterns in the eastern tropical Pacific provide no straightforward information on changes in Atlantic-to-Pacific moisture transport across Central America.

References

Benway HM, Mix AC (2004) Oxygen isotopes, upper-ocean salinity, and precipitation sources in the eastern tropical Pacific. Earth Planet Sci Lett 224:493–507

Benway HM, Mix AC, Haley BA, Klinkhammer GP (2006) Eastern Pacific warm pool paleosalinity and climate variability: 0–30 kyr. Paleoceanography 21:PA3008. doi:10.1029/2005PA001208

Collins WD, Bitz CM, Blackmon ML, Bonan GB, Bretherton CS, Carton JA, Chang P, Doney SC, Hack JJ, Henderson TB, Kiehl JT, Large WG, McKenna DS, Santer BD, Smith RD (2006) The community climate system model version 3 (CCSM3). J Clim 19:2122–2143

Leduc G, Vidal L, Tachikawa K, Rostek F, Sonzogni C, Beaufort L, Bard E (2007) Moisture transport across Central America as a positive feedback on abrupt climate changes. Nature 445:908–911. doi:10.1038/nature05578

Pahnke K, Sachs JP, Keigwin L, Timmermann A, Xie S-P (2007) Eastern tropical Pacific hydrologic changes during the past 27,000 years from D/H ratios in alkenones. Paleoceanography 22:PA4214. doi:10.1029/2007PA001468

Prange M, Steph S, Schulz M, Keigwin LD (2010) Inferring moisture transport across Central America: can modern analogs of climate variability help reconcile paleosalinity records? Quat Sci Rev 29:1317–1321

Romanova V, Prange M, Lohmann G (2004) Stability of the glacial thermohaline circulation and its dependence on the background hydrological cycle. Clim Dyn 22:527–538

Schmittner A, Appenzeller C, Stocker TF (2000) Enhanced Atlantic freshwater export during El Niño. Geophys Res Lett 27:1163–1166

Steph S, Regenberg M, Tiedemann R, Mulitza S, Nürnberg D (2009) Stable isotopes of planktonic foraminifera from tropical Atlantic/Caribbean core-tops: implications for reconstructing upper ocean stratification. Mar Micropaleontol 71:1–19. doi:10.1016/j.marmicro.2008.12.004

Zaucker F, Stocker TF, Broecker WS (1994) Atmospheric freshwater fluxes and their effect on the global thermohaline circulation. J Geophys Res 99:12443–12457

Control of Seasonality and Interannual to Centennial Climate Variability in the Caribbean During the Holocene—Combining Coral Records, Stalagmite Records and Climate Models

Thomas Felis, Denis Scholz, Gerrit Lohmann, Cyril Giry, Claudia Fensterer, Wei Wei and Augusto Mangini

Abstract This study aimed at quantifying the amplitudes of seasonality and interannual to centennial climate variability in the Caribbean region throughout the Holocene, by using marine (shallow-water corals) and terrestrial (speleothems) climate archives, and climate model simulations (COSMOS). Sea-surface temperature (SST) variability on interdecadal to multidecadal timescales was more pronounced during the mid-Holocene compared to the late Holocene. The amplitude of the SST annual cycle was within the present-day range throughout most of the last 6,000 years. Exceptions include slightly increased SST seasonality at 6,200 years ago, which can be attributed mainly to insolation forcing on orbital timescales, and an increased SST seasonality at 2,350 years ago that can be attributed to internal dynamics of the climate system (El Niño-Southern Oscillation). On multidecadal and millennial timescales, precipitation variability during the Holocene was strongly linked to SST in the North Atlantic Ocean, namely the Atlantic Multidecadal Oscillation and variations in the strength of the Atlantic Meridional Overturning Circulation.

T. Felis (✉) · C. Giry
MARUM—Center for Marine Environmental Sciences and Faculty of Geosciences,
University of Bremen, Bremen, Germany
e-mail: tfelis@marum.de

D. Scholz
Institute for Geosciences, Johannes Gutenberg University Mainz, Mainz, Germany

D. Scholz · C. Fensterer · A. Mangini
Heidelberg Academy of Sciences and Humanities, Heidelberg, Germany

G. Lohmann · W. Wei
Alfred Wegener Institute, Helmholtz Centre for Polar and Marine Research,
Bremerhaven, Germany

© The Author(s) 2015
M. Schulz and A. Paul (eds.), *Integrated Analysis of Interglacial Climate Dynamics (INTERDYNAMIC)*, SpringerBriefs in Earth System Sciences,
DOI 10.1007/978-3-319-00693-2_12

Keywords Tropical Atlantic · Caribbean Sea · Holocene · Paleoclimate · Seasonality · Interannual to centennial variability · Corals · Speleothems · Climate models

1 Introduction

Instrumental climate observations during the last 100 years indicate that interactions of the tropical Pacific and Atlantic Oceans play a crucial role in controlling inter-annual to multidecadal climate variability throughout the Caribbean region. Ocean-atmosphere interactions on these timescales play a critical role for regional climate extremes such as droughts, floods and hurricanes. A better understanding of the natural range of variability on these timescales is important for projections of future climate change in this key region of the tropical Atlantic Ocean. However, Caribbean proxy reconstructions that resolve climate variability on these timescales are sparse. We reconstructed and studied the natural range of seasonality and interannual to centennial climate variability in the Caribbean during the Holocene, by using marine and terrestrial climate archives. The major aim was to reconstruct variability of the surface ocean and the atmosphere, and to explain the reconstructed changes with climate model simulations. Shallow-water corals from fossil reef deposits of the southern Caribbean were used to reconstruct temperature and hydrologic balance at the sea surface. Speleothems from caves in the northern Caribbean were used to reconstruct rainfall intensity. Model simulations were used to place the proxy-based results in a large-scale climatic context, and to identify forcing mechanisms of the reconstructed regional climate variability on seasonal to centennial timescales.

2 Materials and Methods

In the southern Caribbean Sea (Bonaire), fossil annually-banded *Diploria strigosa* corals were drilled. Screening for diagenesis indicates the aragonitic coral skeletons are well-preserved (Giry et al. 2010a, 2012). Corals were microsampled at approximately monthly resolution (Giry et al. 2010b) and analyzed for Sr/Ca (proxy for temperature) and $\delta^{18}O$ (proxy reflecting both temperature and seawater $\delta^{18}O$). The records provide monthly resolved reconstructions of sea-surface temperature (SST) and $\delta^{18}O$ seawater (proxy for sea-surface salinity (SSS)) for snapshots since the mid-Holocene. Corals were dated by the $^{230}Th/U$-method and provide a total of ~ 300 years of record. Individual time windows reach up to 68 years in length (Giry et al. 2012, 2013).

In the northern Caribbean Sea (Cuba), stalagmites were recovered from two caves (Dos Anas, Santo Tomas). Three stalagmites were dated by the $^{230}Th/U$-method

(Fensterer et al. 2010, 2012, 2013). Dating uncertainties are between 0.01 and 0.1 thousand years (ka), but are larger (up to 0.8 ka) for a few samples that contain relatively large amounts of detrital Th. Stalagmites were microsampled at a resolution of approximately 2, 4–10, and 15 years, and analyzed for $\delta^{18}O$ and $\delta^{13}C$. The stalagmites grew continuously during the last 1.3 ka and almost continuously during the last 12.5 ka. Stalagmite $\delta^{18}O$ on seasonal to interannual timescales mainly reflects the $\delta^{18}O$ of precipitation, which in turn is mainly controlled by the amount effect (Fensterer et al. 2012, 2013).

Several numerical experiments were performed for three periods of the Holocene using the Earth system model COSMOS (Wei et al. 2012): Pre-industrial [0 ka before present (BP)], mid-Holocene (6 ka BP) and early Holocene (9 ka BP). Configurations with different boundary conditions, i.e., orbital parameters, greenhouse gas concentration, land-sea distribution and freshwater flux from melting of the Laurentide ice sheet (Wei and Lohmann 2012; Wei et al. 2012), allowed to investigate their influence on the climate variability of the Caribbean region during the Holocene on seasonal and interannual to millennial timescales.

3 Key Findings

In the southern Caribbean, monthly resolved coral Sr/Ca and $\delta^{18}O$ records indicate that mid- to late Holocene SST and SSS were characterized by persistent quasi-biennial and prominent interannual to multidecadal variability (Fig. 1) (Giry et al. 2012, 2013). The amplitudes of SST and SSS variations on these timescales have varied since the mid-Holocene. On these timescales, warmer conditions were accompanied by more saline conditions at the sea surface, and vice versa. Potential forcing mechanisms include the wind-induced advection of surface waters from the south and variations in the strength of the Atlantic Meridional Overturning Circulation (AMOC) (Giry et al. 2013). SST variability on inter- to multidecadal timescales was more pronounced during the mid-Holocene compared to the late Holocene and accompanied by enhanced SSS variability (Giry et al. 2012, 2013), a finding that was not detected in the COSMOS simulations. An increased amplitude of the SSS annual cycle and a slightly increased amplitude of the SST annual cycle were reconstructed for the mid-Holocene, which are consistent with the COSMOS simulations (Giry et al. 2012, 2013). These reconstructed changes are mainly attributed to orbitally induced insolation changes that also favor a northward shift of the Intertropical Convergence Zone (ITCZ), resulting in more precipitation in the Caribbean (Fig. 2), especially during summer. Anomalous sea-surface conditions occurred around 2.35 ka BP, characterized by enhanced interannual SST variability at typical El Niño-Southern Oscillation (ENSO) periods, increased amplitude of the SST annual cycle, and a reversal of the SSS annual cycle (Giry et al. 2012, 2013). Such anomalous time intervals were not detected in the COSMOS simulations, which rather suggest a quasi-persistent ENSO influence on the Caribbean during the Holocene (Wei and Lohmann 2013).

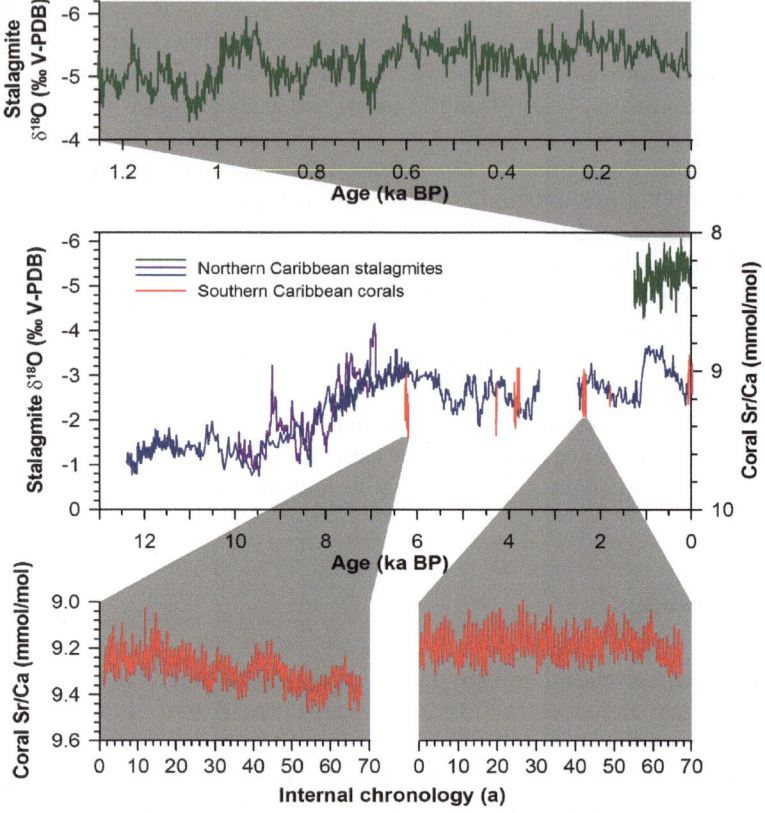

Fig. 1 Biannually to decadally resolved stalagmite $\delta^{18}O$ records from Cuba as a proxy for precipitation in the northern Caribbean Sea. More negative $\delta^{18}O$ values reflect higher precipitation (Fensterer et al. 2012, 2013). Monthly resolved coral Sr/Ca records from Bonaire as a proxy for SST in the southern Caribbean Sea. Lower Sr/Ca ratios reflect higher temperatures (Giry et al. 2012). The two longest records are shown in detail. The corresponding coral $\delta^{18}O$ records and reconstructed $\delta^{18}O$ seawater records (Giry et al. 2013) are not shown

In the northern Caribbean, stalagmite records with a resolution of 4–10 and 15 years indicate a transition from higher to lower $\delta^{18}O$ values between 10 and 6 ka BP (Fig. 1), which is also evident in a planktonic foraminiferal $\delta^{18}O$ record (Fensterer et al. 2013). The amount effect associated with the simulated northward shift of the ITCZ during the early to mid-Holocene cannot fully explain this transition. However, our sensitivity studies using the COSMOS model attribute this transition mainly to the source effect, i.e., to changes in the $\delta^{18}O$ of seawater resulting from melting of the Laurentide ice sheet (Wei and Lohmann 2012). Moreover, a biannually resolved stalagmite $\delta^{18}O$ record indicates that high rainfall amounts on Cuba on multidecadal timescales are strongly related to high SST in the North Atlantic, suggesting an important control of the Atlantic Multidecadal Oscillation (AMO) on northern Caribbean precipitation during the late Holocene

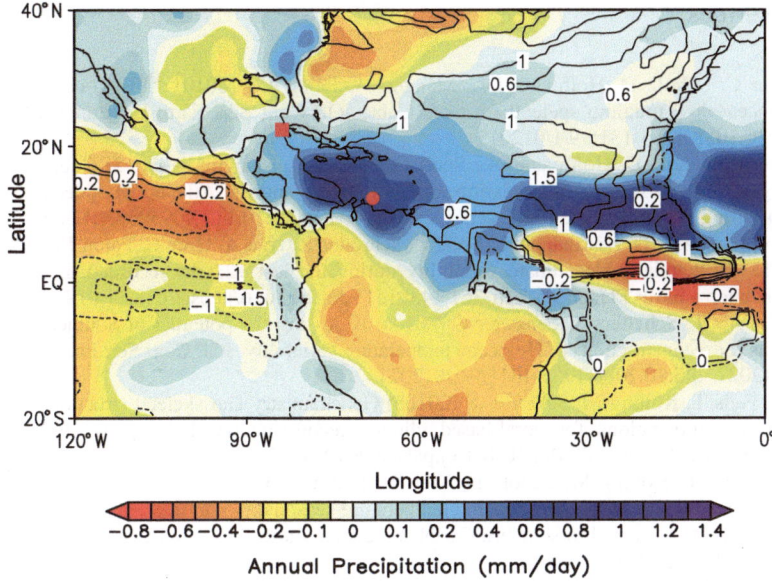

Fig. 2 Simulated annual precipitation (mm/day; *shaded*) and SST seasonality (°C; contour) for the mid-Holocene shown as anomalies relative to the pre-industrial level. Simulations were performed with the Earth system model COSMOS (Wei et al. 2012; Wei and Lohmann 2013). For reference, the stalagmite site at Cuba in the northern Caribbean Sea (*red square*) and the coral site at Bonaire in the southern Caribbean Sea (*red dot*) are shown

(Fensterer et al. 2012). The same relationship is evident on millennial timescales, where North Atlantic cold events coincide with reduced precipitation in Cuba, suggesting a role for the AMOC in controlling northern Caribbean precipitation on these timescales throughout the Holocene (Fensterer et al. 2013).

The COSMOS model simulations, along with southern Caribbean coral-based SST and SSS reconstructions (Giry et al. 2012, 2013) and northern Caribbean stalagmite-based precipitation reconstructions (Fensterer et al. 2012, 2013) indicate that the dominating internal variability in the Caribbean during the Holocene can be linked to ENSO on seasonal and interannual timescales, and to the AMO on multidecadal and longer timescales, with both showing a quasi-persistent feature during the Holocene (Wei and Lohmann 2013). The AMO, however, can be also modulated by background conditions associated with the AMOC. For instance, the large-scale cooling due to the melting flux from the Laurentide ice sheet triggered more vigorous AMOC variations during the early Holocene, and in turn generated a stronger SST signal in the North Atlantic and Caribbean Sea during the AMO warm and cold phase (Wei and Lohmann 2012).

References

Fensterer C, Scholz D, Hoffmann DL, Mangini A, Pajón JM (2010) [230]Th/U-dating of a late Holocene low uranium speleothem from Cuba. IOP Conf Ser: Earth Environ Sci 9:012015. doi:10.1088/1755-1315/9/1/012015

Fensterer C, Scholz D, Hoffmann DL, Spötl C, Pajón JM, Mangini A (2012) Cuban stalagmite suggests relationship between Caribbean precipitation and the Atlantic Multidecadal Oscillation during the past 1.3 ka. Holocene 22:1403–1410. doi:10.1177/0959683612449759

Fensterer C, Scholz D, Hoffmann DL, Spötl C, Schröder-Ritzrau A, Horn C, Pajón JM, Mangini A (2013) Millennial-scale climate variability during the last 12.5 ka recorded in a Caribbean speleothem. Earth Planet Sci Lett 361:143–151. doi:10.1016/j.epsl.2012.11.019

Giry C, Felis T, Scheffers S, Fensterer C (2010a) Assessing the potential of Southern Caribbean corals for reconstructions of Holocene temperature variability. IOP Conf Ser: Earth Environ Sci 9:012021. doi:10.1088/1755-1315/9/1/012021

Giry C, Felis T, Kölling M, Scheffers S (2010b) Geochemistry and skeletal structure of *Diploria strigosa*, implications for coral-based climate reconstruction. Palaeogeogr Palaeoclimatol Palaeoecol 298:378–387. doi:10.1016/j.palaeo.2010.10.022

Giry C, Felis T, Kölling M, Scholz D, Wei W, Lohmann G, Scheffers S (2012) Mid- to late Holocene changes in tropical Atlantic temperature seasonality and interannual to multidecadal variability documented in southern Caribbean corals. Earth Planet Sci Lett 331–332:187–200. doi:10.1016/j.epsl.2012.03.019

Giry C, Felis T, Kölling M, Wei W, Lohmann G, Scheffers S (2013) Controls of Caribbean surface hydrology during the mid- to late Holocene: insights from monthly resolved coral records. Clim Past 9:841–858. doi:10.5194/cp-9-841-2013

Wei W, Lohmann G (2012) Simulated Atlantic Multidecadal Oscillation during the Holocene. J Clim 25:6989–7002. doi:10.1175/JCLI-D-11-00667.1

Wei W, Lohmann G (2013) Simulated Caribbean climate variability during the mid-Holocene. In: Lohmann G, Grosfeld K, Wolf-Gladrow D, Unnithan V, Notholt J, Wegner A (eds) Earth system science: bridging the gaps between disciplines. Perspectives from a multi-disciplinary Helmholtz Research School, SpringerBriefs in Earth System Sciences. Springer, Heidelberg, pp 64–69. doi:10.1007/978-3-642-32235-8

Wei W, Lohmann G, Dima M (2012) Distinct modes of internal variability in the global meridional overturning circulation associated with the Southern Hemisphere westerly winds. J Phys Oceanogr 42:785–801. doi:10.1175/JPO-D-11-038.1

The Southern Westerlies During the Holocene: Paleoenvironmental Reconstructions from Chilean Lake, Fjord, and Ocean Margin Sediments Combined with Climate Modeling

Frank Lamy, Matthias Prange, Helge W. Arz, Vidya Varma,
Jerome Kaiser, Rolf Kilian, Jens Hefter, Albert Benthien
and Gesine Mollenhauer

Abstract This project aimed at investigating centennial to millennial-scale changes of the strength and position of the southern westerly wind belt (SWW) using multi-proxy paleoprecipitation and paleoceanographic records combined with transient model runs. The proxy data records reveal a distinct latitudinal anti-phasing of wind changes between the core and northern margin of the SWW over the Holocene. During the early Holocene, the SWW core was enhanced and the northern margin was reduced, whereas the opposite pattern is observed in the late Holocene. These Holocene changes resemble modern seasonal wind belt variations and can be best explained by varying sea-surface temperature fields in the eastern South Pacific. Transient modeling experiments from the mid- to late Holocene are not yet consistent with these proxy results. However, a good data-model agreement exists when investigating the potential impact of solar variability on the SWW at centennial time-scales during the latest Holocene with periods of lower (higher) solar activity causing equatorward (southward) shifts of the SWW.

F. Lamy (✉) · J. Hefter · A. Benthien · G. Mollenhauer
Alfred Wegener Institute, Helmholtz Centre for Polar and Marine Research,
Bremerhaven, Germany
e-mail: frank.lamy@awi.de

M. Prange · V. Varma
MARUM—Center for Marine Environmental Sciences and Faculty of Geosciences,
University of Bremen, Bremen, Germany

H.W. Arz · J. Kaiser
Leibniz Institute for Baltic Sea Research Warnemünde, Rostock-Warnemünde, Germany

R. Kilian
Department of Geology, Faculty of Geography/Geosciences, University of Trier,
Trier, Germany

© The Author(s) 2015
M. Schulz and A. Paul (eds.), *Integrated Analysis of Interglacial Climate Dynamics (INTERDYNAMIC)*, SpringerBriefs in Earth System Sciences,
DOI 10.1007/978-3-319-00693-2_13

75

Keywords Paleoclimatology · Paleoceanography · Climate modeling · Holocene ·
Solar variability · Westerly winds

1 Introduction

The westerlies are major zonal atmospheric circulation systems in both northern and
southern hemispheres. Particularly, the southern westerly wind belt (SWW) exerts a
strong control on global climate and oceanography. On a hemispheric scale, SWW
changes contribute substantially to the forcing of the deep and vigorous Antarctic
Circumpolar Current, while wind-induced upwelling in the Southern Ocean raises
large amounts of deep water to the surface affecting the global thermohaline circulation and atmospheric CO_2 contents. Observational data suggest that the
southern margin of the SWW has intensified over the past 40 years, a trend that is
expected to proceed over the next centuries. This may provide a positive feedback
on global warming through reducing the uptake of anthropogenic CO_2 or even
promoting outgassing of old naturally stored CO_2 through upwelling.

Southern South America is the only landmass intersecting both the present core
(southernmost Patagonia) and the northern margin (central Chile) of the wind-belt.
Since the SWW nearly entirely control precipitation on the western side of the
southern Andes, proxy records of past precipitation changes from this region are
ideal to reconstruct past variability of the SWW. Furthermore, the ocean margin,
lake, and fjord sediments provide high resolution paleoclimate archives along the
southern Chilean continental margin suitable to reconstruct precipitation changes
and related ocean variations during the Holocene (e.g., Kilian and Lamy 2012).
These sediment records combined with climate modeling can be used to place the
short instrumental data-sets into a longer-term perspective covering centennial-
millennial-scale SWW changes during the complete Holocene.

2 Materials and Methods

Our primary goal is to integrate southern Chilean lake, fjord, and ocean margin
proxy archives with climate modeling in order to reconstruct Holocene changes
in vegetation, rainfall, and lake/sea-surface temperatures and relate those to changes
in the strength and latitudinal position of the SWW. Therefore, we applied a number
of different methods including non-destructive core-logging methods to obtain
ultra-high-resolution records, sedimentological, and palynological approaches
(Lamy et al. 2010). Large efforts were devoted to generate detailed and reliable age
models by radiocarbon dating and tephrostratigraphy. Furthermore, we investigated
the regional calibration and applicability of novel organic biomarkers including
glyceryl dialkyl glycerol tetraether (GDGT)—based proxies (such as TEX86 to

reconstruct water temperature and the MBT/CBT index to reconstruct mean air temperature) as well as compound-specific hydrogen isotopic composition of lipids derived from land-plant leafwaxes (e.g., long-chain, odd-numbered n-alkanes) to reconstruct the hydrological cycle. Application of these molecular proxies on a Holocene fjord sediment record from the center of the SWW and covering the Holocene shows promising results.

Numerical transient experiments using the comprehensive global climate model CCSM3 (Community Climate System Model version 3) were carried out in order to simulate the evolution of the SWW under orbital forcing from the mid-Holocene [7,000 years (ka) before present (BP)] to pre-industrial modern times (250 years BP). These simulations were accompanied by a model inter-comparison with orbitally forced Holocene transient simulations from four other coupled global climate models (Varma et al. 2012a). In order to study the response of SWW to solar variability, model runs with idealized solar forcing were performed (Varma et al. 2011). Finally, we investigated the influence of the stratosphere and its ozone content on SWW variability, using two transient simulations (one with fixed and one with solar-induced varying stratospheric ozone) with the coupled atmosphere-ocean general circulation model EGMAM (ECHO-G with Middle Atmosphere Model) focusing on the periods of the Late Maunder Minimum (LMM: 1675–1715 AD) and Pre-Industrial (PI: 1716–1790 AD) (Varma et al. 2012b).

3 Key Findings

Our multi-proxy compilation based on fjord and lake sediment records from the hyperhumid zone of southernmost Chile provides a consistent picture of Holocene SWW variability (Fig. 1; Lamy et al. 2010). Precipitation and thus SWW strength changes deduced from humidity sensitive pollen, precipitation-dependent terrestrial organic carbon accumulation in lake and fjord sediments (Fig. 1e), and salinity dependent decrease in biogenic carbonate accumulation indicate wetter/windier conditions between ∼12.5 and ∼8.5 ka BP, intermediate conditions thereafter until ∼5.5 ka BP, and finally reduced precipitation and less intense westerlies during the late Holocene (Fig. 1). At the northern margin of the SWW in central Chile reconstructed rainfall changes are generally anti-phased to those from the core zone. A rainfall reconstruction based on Lake Aculeo level changes (34°S; Fig. 1a; Jenny et al. 2003) reveals substantially reduced precipitation during the early and mid-Holocene in agreement with reduced terrigenous sediment input in Lake LleuLleu (37°S; Fig. 1b) and increased input of Andean-derived versus coastal-derived terrestrial material at the continental margin (41°S; Fig. 1c; Lamy et al. 2001). All three records show a pronounced shift to more humid conditions starting at ca. 5.5 ka BP and extending throughout the late Holocene. Though these records only reach back to ∼8–10 ka BP, substantially lower rainfall during the early Holocene and thus reduced westerly influence has been reconstructed at many sites in the region (Latorre et al. 2007).

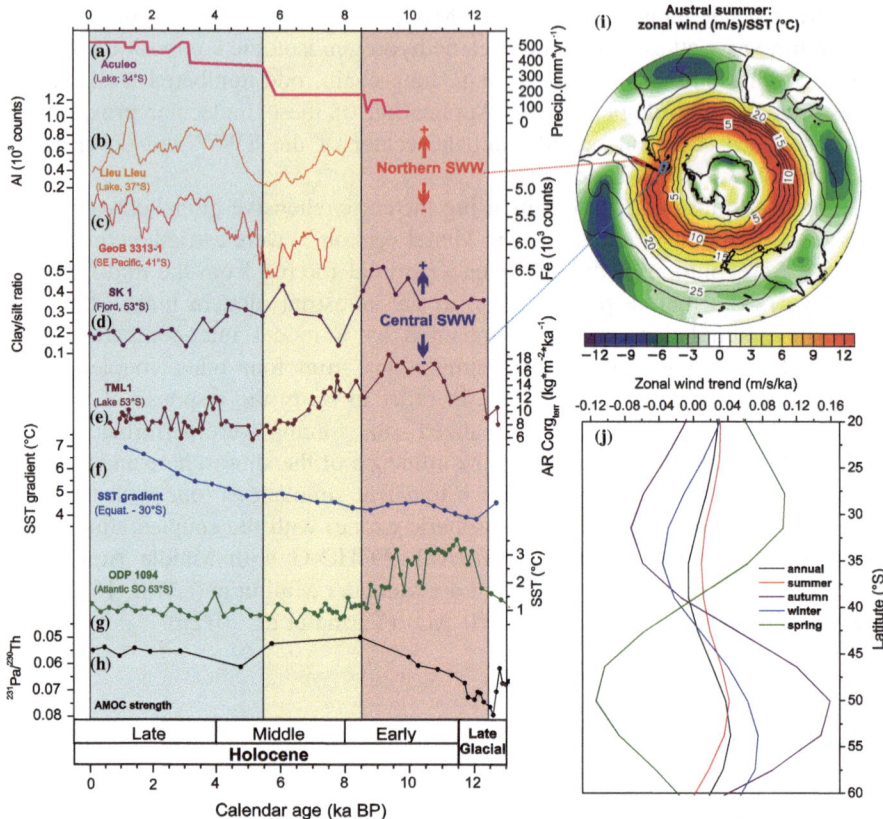

Fig. 1 Proxy records for Holocene changes of the SWW compared to transient model results. **a** Precipitation reconstruction from Lake Aculeo, central Chile (Jenny et al. 2003). **b** Al content record from core LL-KL009, Lake LleuLleu, central Chile as a paleoprecipitation proxy. **c** Fe content changes recorded in core GeoB 3313-1 retrieved from the Chilean continental margin at 41°S. **d** Silt/clay ratio record from core SK1 (fjord Seno Skyring) recording the long-distance eastward transport of illite-rich Andean clay from the Patagonian Batholith, which can be taken as a "direct" wind strength proxy (Lamy et al. 2010). **e** Terrestrial organic carbon accumulation rate record from lake core TML1 in the Strait of Magellan fjord region (Lamy et al. 2010). **f** SST gradients between the eastern tropical Pacific (core V19-28; Koutavas and Sachs 2008) and a mid-latitude SE Pacific record (core GeoB 7139; Kaiser et al. 2008). **g** Diatom assemblage-based summer SST record from the Atlantic sector of the Southern Ocean at ~53°S (Bianchi and Gersonde 2004). **h** ^{231}Pa/^{230}Th record from a subtropical North Atlantic sediment core taken as a proxy for the strength of the Atlantic meridional overturning circulation (McManus et al. 2004). *Vertical bars* mark the multi-millennial Holocene periods (*red* = early Holocene intensification of the core SWW and weakening of the northern margin; *blue* = late Holocene weakening of the core SWW and intensification of the northern margin; *yellow* = intermediate conditions during the middle Holocene). **i** Southern hemisphere zonal wind (m/s) climatology at 850 hPa and SSTs for austral summer. Study areas in central and southern Chile are marked. **j** Zonally averaged seasonal and annual trends in low-level zonal wind for the period 7 ka BP to pre-industrial. Results are from CCSM3 (Varma et al. 2012a)

The early Holocene SWW core maximum coincides with a widespread warming at southern hemisphere mid-latitudes as evidenced by sea-surface temperature (SST) records from the Atlantic Southern Ocean (53°S; Fig. 1g) (Bianchi and Gersonde 2004) as well as offshore southern (Caniupan et al. 2011), central (Lamy et al. 2002) and northern Chile (Kaiser et al. 2008). At the same time, SSTs in the eastern tropical Pacific were relatively cold (Koutavas and Sachs 2008) reducing the low to mid-latitude SST gradient in the South Pacific (Fig. 1f) and consequently the SWW at their northern margin as during present-day summer. Warming in the eastern tropical Pacific and cooling further south during the late Holocene possibly enhanced the latitudinal SST gradients resulting in stronger winds at the northern margin of the SWW (Lamy et al. 2010) and reduced winds across the southern tip of South America as presently occurring during winter.

Some of our proxy data are inconsistent with our numerical model results which suggest that the annual and seasonal mean SWW is subjected to an overall strengthening and poleward shifting trend during the course of the mid-to-late Holocene under the influence of orbital forcing, except for the austral spring season, where the SWW exhibit an opposite trend of shifting towards the equator (Fig. 1j) (Varma et al. 2012a). The major change in the proxy data occurs during the early Holocene (not yet covered by the transient model runs) rather than the mid- and late Holocene. However, the modeled poleward shifting and strengthening of the westerlies during most of the year from the mid to the late Holocene is inconsistent with the trend to more humid conditions and stronger northern margin westerlies in central Chile shown by the data (Fig. 1a–c). An exception is the modeled austral spring enhancement of the northern westerlies, which is however not the major rainfall season in central Chile assumed to be recorded by the proxies (Fig. 1j). Further work is therefore needed to reconcile these model-data inconsistencies including an extension of the transient experiments into the early Holocene where processes such as the ocean-driven bipolar see-saw become important (Fig. 1h).

More consistent data-model results have been obtained for centennial-scale SWW changes in central Chile over the past 3 ka (Fig. 2) (Varma et al. 2011, Varma et al. 2012b). The proxy and model results suggest that centennial-scale periods of lower (higher) solar activity caused equatorward (southward) shifts of the annual mean SWW. Under a "bottom-up" mechanism, where applied changes in total solar irradiance mostly affect the climate system through shortwave absorption by the surface, the strength and position of the SWW are strongly related to meridional surface temperature gradients. By contrast, a "top-down" mechanism influences the troposphere via stratospheric ozone responses to variations in ultraviolet radiation and dynamical coupling between the atmospheric layers. The SWW response in simulations with varying stratospheric ozone (EGMAM2) is more pronounced and robust compared to the one with fixed ozone (EGMAM1) (Fig. 2) suggesting an important contribution from the middle atmosphere through a "top-down" mechanism.

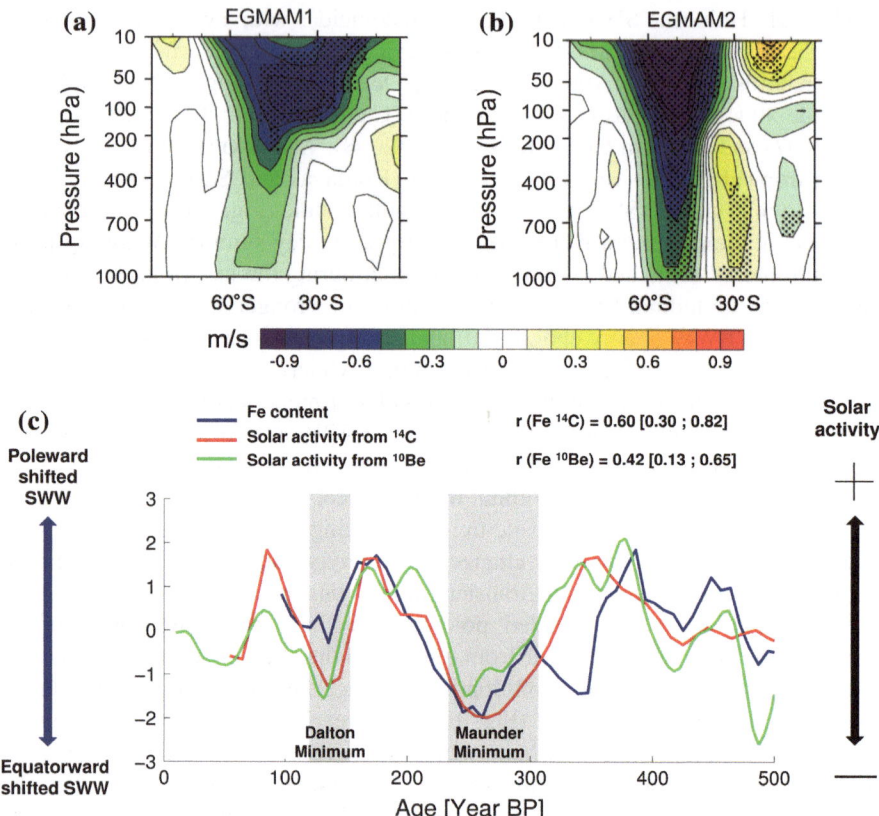

Fig. 2 Annual-mean zonal wind anomalies (LMM-mean minus PI-mean) in the EGMAM simulations along with the reconstructions for SWW position and solar activity. **a** and **b** Zonally averaged zonal wind anomalies in the Southern Hemisphere in EGMAM1 (fixed ozone) and EGMAM2 (varying ozone) simulations respectively. Stippling indicates significance of the anomaly at the 0.05 level according to a Student's t-test. **c** Reconstruction of the SWW position [*blue line*, based on the GeoB3313-1 iron record (Lamy et al. 2001)] versus solar activity based on ^{10}Be (*green line* Steinhilber et al. 2009) and ^{14}C (*red line* Solanki et al. 2004), for the last 500 years. *Grey bars* mark the low solar activity periods of the Dalton minimum and the Maunder minimum. Time series are unsmoothed, detrended and standardized. The negative (positive) iron anomalies suggest northward (southward) shifted SWW (Lamy et al. 2001; Varma et al. 2011). 95 % confidence intervals (in brackets) for Pearson correlation coefficients (r) were calculated using a bootstrap method, where autocorrelation has been taken into account

References

Bianchi C, Gersonde R (2004) Climate evolution at the last deglaciation: the role of the southern ocean. Earth Planet Sci Lett 228:407–424

Caniupan M, Lamy F, Lange CB, Kaiser J, Arz H, Kilian R, Urrea OB, Aracena C, Hebbeln D, Kissel C, Laj C, Mollenhauer G, Tiedemann R (2011) Millennial-scale sea surface temperature and patagonian ice sheet changes off southernmost chile (53°S) over the past similar to 60 kyr. Paleoceanography 26(3)

Jenny B, Wilhelm D, Valero-Garces BL (2003) The southern westerlies in central chile: Holocene precipitation estimates based on a water balance model for laguna aculeo (33°50' S). Clim Dyn 20:269–280

Kaiser J, Schefuss E, Lamy F, Mohtadi M, Hebbeln D (2008) Glacial to Holocene changes in sea surface temperature and coastal vegetation in north central chile: high versus low latitude forcing. Quat Sci Rev 27:2064–2075

Kilian R, Lamy F (2012) A review of glacial and Holocene paleoclimate records from southernmost patagonia (49–55°S). Quat Sci Rev 53:1–23

Koutavas A, Sachs JP (2008) Northern timing of deglaciation in the eastern equatorial pacific from alkenone paleothermometry. Paleoceanography 23:PA4205. doi:10.1029/2008PA001593

Lamy F, Hebbeln D, Rohl U, Wefer G (2001) Holocene rainfall variability in southern chile: a marine record of latitudinal shifts of the southern westerlies. Earth Planet Sci Lett 185:369–382

Lamy F, Rühlemann C, Hebbeln D, Wefer G (2002) High- and low-latitude climate control on the position of the southern peru-chile current during the Holocene. Paleoceanography 17(2):1028. doi:10.1029/2001PA000727

Lamy F, Kilian R, Arz HW, Francois JP, Kaiser J, Prange M, Steinke T (2010) Holocene changes in the position and intensity of the southern westerly wind belt. Nat Geosci 3:695–699

Latorre C, Moreno PI, Vargas G, Maldonado A, Villa-Martínez R, Armesto JJ, Villagrán C, Pino M, Núñez L, Grosjean M (2007) Late quaternary environments and paleoclimate. In: Moreno T, Gibbons W (eds) The geology of chile. The London Geological Society Press, London, pp 309–328

McManus J, Francois R, Gherardi J-M, Keigwin LD, Brown-Leger S (2004) Collapse and rapid resumption of Atlantic meridional circulation linked to deglacial climate changes. Nature 428:834–837

Solanki SK, Usoskin IG, Kromer B, Schussler M, Beer J (2004) Unusual activity of the Sun during recent decades compared to the previous 11,000 years. Nature 431:1084–1087

Steinhilber F, Beer J, Fröhlich C (2009) Total solar irradiance during the Holocene. Geophys Res Lett. doi:10.1029/2009GL040142

Varma V, Prange M, Lamy F, Merkel U, Schulz M (2011) Solar-forced shifts of the southern hemisphere westerlies during the Holocene. Clim Past 7:339–347

Varma V, Prange M, Merkel U, Kleinen T, Lohmann G, Pfeiffer M, Renssen H, Wagner A, Wagner S, Schulz M (2012a) Holocene evolution of the southern hemisphere westerly winds in transient simulations with global climate models. Clim Past 8:391–402

Varma V, Prange M, Spangehl T, Lamy F, Cubasch U, Schulz M (2012b) Impact of solar-induced stratospheric ozone decline on southern hemisphere westerlies during the late maunder minimum. Geophys Res Lett. doi:10.1029/2012GL053403

Mineral Dust Variability in Antarctic Ice for Different Climate Conditions

Anna Wegner, Natalia Sudarchikova, Hubertus Fischer and Uwe Mikolajewicz

Abstract This study aims to understand the dust deposition changes on the Antarctic ice sheet in different climatic stages. To this end, high-resolution dust concentration and size profiles from the EPICA-DML ice core over the transition from the last glacial to the Holocene (T1) were combined with model experiments for four interglacial time slices and the Last Glacial Maximum (LGM). A strong decrease in dust concentration (factor 46) and a slight increase in dust size was observed during T1. A strong coupling between transport and intensified sources during the glacial could be derived from the seasonal variability of concentration and size and its phase-lag. This strong coupling vanishes during the Holocene. The model simulates increased dust deposition in Antarctica for all past interglacial time slices compared to the pre-industrial period. The major cause for the increase is enhanced Southern Hemisphere dust emission, but changes in atmospheric transport are also relevant. The maximum dust deposition in Antarctica is simulated for the LGM, showing a 10-fold increase compared to preindustrial conditions.

Keywords Mineral dust · Antarctica · Glacial and interglacial periods · Climate model · Ice core · Atmospheric transport

A. Wegner (✉)
Helmholtz Centre for Polar and Marine Research, Alfred Wegener Institute, Bremerhaven, Germany
e-mail: anna.wegner@awi.de

N. Sudarchikova · U. Mikolajewicz
Max Planck Institute for Meteorology, Hamburg, Germany

N. Sudarchikova
Climate Service Center and Helmholtz-Zentrum Geesthacht—Centre for Materials and Coastal Research, Hamburg, Germany

H. Fischer
Climate and Environmental Physics, Physics Institute and Oeschger Centre for Climate Change Research, University of Bern, Bern, Switzerland

© The Author(s) 2015
M. Schulz and A. Paul (eds.), *Integrated Analysis of Interglacial Climate Dynamics (INTERDYNAMIC)*, SpringerBriefs in Earth System Sciences, DOI 10.1007/978-3-319-00693-2_14

1 Introduction

Polar ice cores represent a unique archive for the deposition of aeolian dust particles in the past, as mineral dust was transported over long distances from desert regions to the polar ice sheets (e.g., Lambert et al. 2008) and is less influenced by local atmospheric conditions then other archives. While the total dust deposition is a first order measure of dust mobilization, hence, climate conditions in the dust source region (Fischer et al. 2007a), particle-size distribution is influenced by transport efficiency.

The overall goal of this study is a quantitative interpretation of Antarctic ice core dust records from the inception to the end of interglacial periods in terms of changes in dust mobilization and transport. Here we use ice-core dust concentration, size and chemical composition as well as model analysis for selected time slices and combine the two in order to assess the dust input to the Antarctic ice sheet quantitatively. This yields information about the emissions in the dust source regions as well as about changes in atmospheric circulation patterns responsible for dust transport to Antarctica on time scales ranging from seasonal to stadial-to-interstadial. In this project variability on these timescales and their causes are investigated.

2 Materials and Methods

Using the EPICA Dronning Maud Land (EDML) ice core, continuous profiles of dust concentration and size were obtained using a Laser Particle Counter (LPD, Klotz Company Bad Liebenzell) and evaluated to provide a full picture of dust transport changes over the transition from the last glacial to the Holocene (T1). Generally, the LPD uses scattering and shadowing of laser light as detection method, which is calibrated using spherical latex particles. Thus, the analysis of non-spherical dust particles in ice cores could be affected by shape artefacts. Here we performed an additional calibration, where we compared the LPD results with Coulter Counter data, which measures the size as volume directly, from the same depth intervals and corrected for the shape. In previous studies only non sea salt calcium (nss-Ca) was used as dust proxy in the EDML ice core (Fischer et al. 2007b), which represents the soluble fraction of the dust. Here, we analysed particulate dust, representing the insoluble fraction of the dust. The main advantage of the particulate dust is the possibility to obtain additionally the dust size, as an indicator for the transport intensity.

The atmospheric general circulation model with online coupled interactive dust scheme ECHAM5/HAM (Stier et al. 2005) was used to study the dust cycle for the interglacial time slices 6 thousand years (ka) BP (before present, where present is defined as 1950, mid-Holocene), 126 ka BP (Eemian) and 115 ka BP (last glacial inception). Additionally, a pre-industrial control simulation (CTRL) was performed. The glacial time slice 21 ka BP (Last Glacial Maximum (LGM)) was simulated as well. The model resolution was T31 (approx. $3.75° \times 3.75°$). Sea-surface temperature (SST) and sea ice distribution were taken from equilibrium simulations with the coupled atmosphere-ocean model ECHAM5/MPIOM (Mikolajewicz et al. 2007). The setup followed the PMIP2 protocol (Braconnot et al. 2007). For time slices 115 and 126 ka BP insolation was changed accordingly, greenhouse gas concentrations were kept at 6 ka BP level. The vegetation distribution was specified according to simulations with the dynamical vegetation model LPJ (Smith et al. 2001) forced with output from the coupled model. A land grid point was defined as a potential dust source, if the maximal vegetation cover was below 25 %. A detailed description of the model setup can be found in Sudarchikova (2012).

3 Key Findings

The measured dust concentration decreases at EDML over T1 by a factor of 46, i.e., in a similar range as in Dome C and at Vostok (Lambert et al. 2008; Petit et al. 1999). The previously used non nss-Ca record (Fig. 1, Fischer et al. 2007b) yields slightly lower values due to analytical uncertainties in the Holocene samples. An absolute minimum of the dust concentration occurs at ~ 11 ka BP, which is in line with other records from the east Antarctic plateau (EAP). Larger sizes (~ 2.3 μm) were found during the Holocene compared to the glacial (~ 2 μm). This is in line with observations for Dome C (Delmonte et al. 2002). At EDML an absolute minimum in the dust size occurs at 16 ka BP. This can also be identified in the EPICA Dome C ice core (EDC), but due to the seasonal resolution in the EDML core, which is not present in EDC, it is more pronounced in EDML. During the Holocene and until MIS 3 seasonal signals in the dust concentration and size can be observed, with amplitudes of the dust concentration up to a factor of 30.

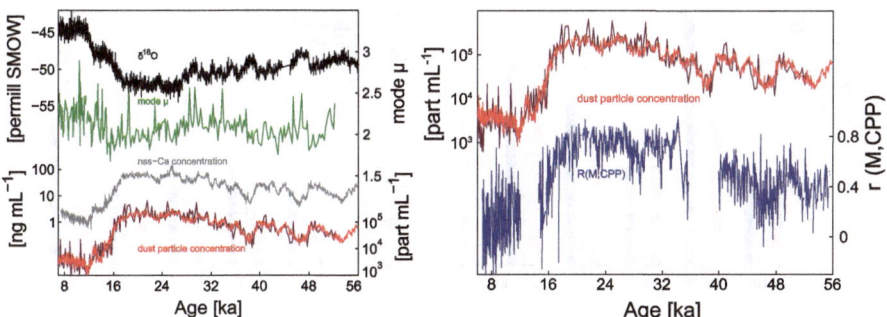

Fig. 1 Dust profile over T1 in the EDML ice core. Oxygen Isotopes, dust concentration (M) and size (as coarse particle percentage (CPP: mass in the size range 3–5 μm divided by the mass concentration in the size range 1–5 μm))and nss-Ca concentrations (Fischer et al. 2007b) *(left)*, dust concentration and phasing of dust concentration and size *(right)*. For the phasing the correlation coefficient (r) between the dust concentration M and the size (CPP) was calculated over 1 m intervals each

During the glacial a clear correlation between the seasonal cycles of dust concentration and size can be found (r = 0.8). During T1, beginning from 19 ka BP, the correlation decreases to values below 0.4 in the Holocene.

For all time slices the simulated dust deposition in Antarctica is increased relative to the pre-industrial CTRL (Fig. 2b). In the mid-Holocene, dust deposition is increased by a factor of 3.8, and in the Eemian by a factor of 2.7. Dust deposition in the last glacial inception is only slightly enhanced. The highest dust deposition in Antarctica is simulated for the LGM, showing a 10.2-fold increase compared to CTRL.

The modeled increase in dust concentration (dust deposition rate divided by precipitation) fits to the results from ice cores for 115 and 126 ka BP (see Fig. 2a). However the model overestimates the 6 ka BP to pre-industrial ratio substantially likely due to overestimation of the South American and Australian dust sources. The prescribed vegetation is the main source of uncertainty in our model simulations of the dust cycle for past time slices. The dust concentration in Antarctic ice for the LGM is further enhanced by the strongly reduced precipitation.

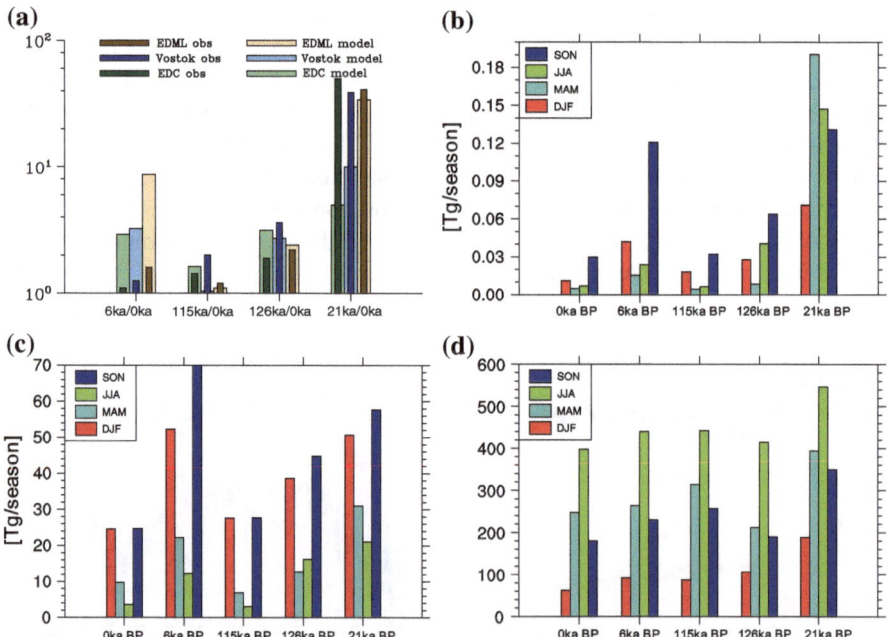

Fig. 2 **a** Ratios of dust mass concentration in the Antarctic ice for model simulations and observations for 6, 115, 126 ka BP and 21 ka BP with respect to pre-industrial. The records shown in dark colours represent Coulter Counter data from the literature. Data from the closest model grid point are shown in light colours. **b** Integrated modelled dust deposition in Antarctica [Tg/season]; **c** Modelled southern hemisphere dust emissions [Tg/season]; **d** Number of trajectories/season originating at 500 and 800 hPa from dust source grid boxes and reaching Antarctica within 10 days

Approximately two thirds of the increase in dust deposition over Antarctica for the mid-Holocene and Eemian is attributed to enhanced Southern Hemisphere dust emissions (Fig. 2c), predominantly from the Australian source (not shown). Atmospheric transport efficiency (described by the number of air mass trajectories originating above the Southern Hemisphere grid points with dust emissions and reaching Antarctica within 10 days) is shown in Fig. 2d. Slightly increased transport efficiency in 6 and 126 ka BP causes the remaining one third of the increase in dust deposition in Antarctica. In general, the annual cycle of emission and the efficiency of the atmosphere to transport dust particles to Antarctica are out of phase for the considered time slices. In the LGM simulation, dust deposition over Antarctica is significantly increased due to 2.6 times higher Southern Hemisphere dust emissions, doubled atmospheric transport efficiency and 30 % weaker precipitation over the Southern Ocean (not shown).

The ice core reconstruction cannot pinpoint the absolute season of dust input to the Antarctic ice during the LGM, but it can pinpoint the onset of the decoupling of transport and emission. A shift between the maxima of dust concentration and size suggests that a different seasonality of dust emission and transport intensity may have contributed to the dust variability over T1. The different phasing of intense transport and high dust concentration in the ice starts very early during T1 (∼ 19 ka BP, Fig. 1) simultaneously with the onset of the decrease in dust concentration. The changes in the seasonality of dust concentration in the ice obtained by the model support this finding.

According to the ice cores, for the whole EAP the same dust provenance (southern South America) is dominant during the glacial, whereas during the Holocene different source regions could be detected in different areas of the EAP. For the Indian sector of the EAP Australia plays a major role. For the Atlantic sector some contributions from other sources outside the South American continent could be detected. However, the major contribution originates from southern South America. The South American dust sources are dominant until 15 ka BP. From that time on contributions from sources outside South America can be detected (Wegner et al. 2012). At that time the dust concentration in Antarctic ice cores are almost on Holocene levels. The model suggests a major influence of South America to EDML during the Holocene (not shown).

References

Braconnot P, Otto-Bliesner B, Harrison S, Joussaume S, Peterschmitt JY, Abe-Ouchi A, Crucifix M, Driesschaert E, Fichefet T, Hewitt CD, Kageyama M, Kitoh A, Laîné A, Loutre MF, Marti O, Merkel U, Ramstein G, Valdes P, Weber SL, Yu Y, Zhao Y (2007) Results of PMIP2 coupled simulations of the Mid-Holocene and Last Glacial Maximum—Part 1: experiments and large-scale features. Clim Past 3:261–277

Delmonte B, Petit JR, Maggi V (2002) LGM-Holocene changes and Holocene millenial scale oscillations of dust particles in the EPICA Dome C ice core, East Antarctica. Ann Glaciol 35:306–3012

Fischer H, Siggaard-Andersen ML, Ruth U, Roethlisberger R, Wolff E (2007a) Glacial/Interglacial changed in mineral dust and sea-salt records in polar ice cores: sources, transport, and deposition. Rev Geophys 45

Fischer H, Fundel F, Ruth U, Twarloh B, Wegner A, Udisti R, Becagli S, Castellano E, Morganti A, Severi M, Wolff E, Littot G, Roethlisberger R, Mulvaney R, Hutterli MA, Kaufmann P, Federer U, Lambert F, Bigler M, Hansson M, Jonsell U, de Angelis M, Boutron C, Siggaard-Andersen ML, Steffensen JP, Barbante C, Gaspari V, Gabrielli P, Wagenbach D (2007b) Reconstruction of millenial changes in the dust emission, transport and regional sea ice coverage using the deep EPICA ice cores from the Atlantic and the Indian sector of Antarctica. Earth Planet Sci Lett 260:340–354. doi:10.1016/j.epsl.2007.06.014

Lambert F, Delmonte B, Petit JR, Bigler M, Kaufmann PR, Hutterli MA, Stocker TF, Ruth U, Steffensen JP, Maggi V (2008) Dust-climate couplings over the past 800,000 years from the EPICA Dome C ice core. Nature 452:616–619

Mikolajewicz U, Vizcaino M, Jungclaus J, Schurgers G (2007) Effect of ice sheet interactions in anthropogenic climate change simulations. Geophys Res Lett 34:L18706

Petit JR, Jouzel J, Raynaud D, Barkov NI, Barnola JM, Basile I, Bender M, Chappellaz J, Davisk M, Delaygue G, Delmotte M, Kotlyakov VM, Legrand M, Lipenkov VY, Lorius C, Pepin L, Ritz C, Saltzmank E, Stievenard M (1999) Climate and atmospheric history of the past 420,000 years from the Vostok ice core, Antarctica. Nature 399:429–436

Smith B, Prentice C, Sykes MT (2001) Representation of vegetation dynamics in the modelling of terrestrial ecosystems: comparing two contrasting approaches within European climate space. Global Ecol Biogeogr 10:621–637

Stier P, Feichter J, Kinne S, Kloster S, Vignati E, Wilson J, Ganzeveld L, Tegen I, Werner M, Balkanski Y, Schulz M, Boucher O, Minikin A, Petzold A (2005) The aerosol-climate model ECHAM5-HAM. Atmos Chem Phys 5:1125–1156

Sudarchikova N (2012) Modeling of mineral dust in the Southern Hemisphere with focus on Antarctica for interglacial and glacial climate conditions. Dissertation, University of Hamburg, Germany

Wegner A, Gabrielli P, Dick D, Ruth U, Kriews M, De Deckker P, Barbante C, Cozzi G, Delmonte B, Fischer H (2012) Change in dust variability in the Atlantic sector of Antarctica at the end of the last deglaciation. Clim Past 8:135–147

Model-Data Synthesis of Monsoon Amplitudes for the Holocene and Eemian

Birgit Schneider, Ralph R. Schneider, Yiming Wang
and Vyacheslav Khon

Abstract Monsoon intensity is driven by changes in hemispheric summer insolation. Marine proxy data show distinct glacial-interglacial variability with changes in vegetation and weathering inferred from the terrigenous fraction, e.g., by plant lipid and mineral composition. Unfortunately, no quantitative evidence is available for differences in monsoonal precipitation. A sensitivity study with a vegetation model implies that C_4/C_3 ratios are influenced by individual changes in precipitation, CO_2, and temperature. Therefore, sedimentary $\delta^{13}C$ records of land plant lipids are no unambiguous indicator for humidity-driven changes in paleo-vegetation. A novel indicator of past humidity over continents, the δD signature of leaf waxes, suggests similar conditions for the Indian summer monsoon during the Holocene and Eemian. However, this new proxy requires more detailed regional studies, since climate model simulations clearly show significant differences in monsoon strength between interglacial periods. Accordingly, a more intense hydrological cycle is expected for the Eemian due to an overall warmer climate driven by precessional forcing.

Keywords Paleomonsoon · Precipitation reconstructions · Marine and terrestrial proxy data · Climate modeling · Coupled atmosphere-ocean hydrological cycle

1 Introduction

The frequencies of climate extremes associated with the hydrological cycle in low latitudes have increased during the last decades due to global warming (IPCC 2007). It is important to know how the hydrological cycle responds to global warming, because intra- and interannual changes associated with monsoon rainfall have

B. Schneider (✉) · R.R. Schneider · Y. Wang · V. Khon
Department of Geosciences, Institute of Geosciences, University of Kiel,
Kiel, Germany
e-mail: bschneider@gpi.uni-kiel.de

© The Author(s) 2015   89
M. Schulz and A. Paul (eds.), *Integrated Analysis of Interglacial Climate
Dynamics (INTERDYNAMIC)*, SpringerBriefs in Earth System Sciences,
DOI 10.1007/978-3-319-00693-2_15

important consequences for the livelihood of billions of people. At orbital time scales, the most prominent factor controlling the monsoon is local summer insolation, as indicated by fluctuations of monsoon strength coinciding remarkably well with changes in the Earth's precessional cycle (Cruz et al. 2005; Wang et al. 2001, 2008). Moreover, records of monsoon variability derived from paleoclimate proxy data suggest increased monsoon intensity during warm interglacials and shorter term interstadials compared to cold glacials and stadials (i.e., Cruz et al. 2005, 2009; Fleitmann et al. 2003; Wang et al. 2001, 2008). The climate of the Eemian ($\sim$ 130 thousand years (ka) before present (BP)—115 ka BP) was probably by 1–2 °C warmer than that of the Holocene (Kukla et al. 2002; Leduc et al. 2010), so that an intensified hydrological cycle and thus monsoon can be expected due to the temperature dependence of the atmospheric water holding capacity (see e.g., Cruz et al. 2005). Although climate model results show more intense monsoon systems under warmer Eemian climate (Braconnot et al. 2008; Kutzbach et al. 2008), from paleoproxy records no clear evidence is available for differences in monsoon intensity between different interglacials or interstadials. Therefore, in the present study a new paleoprecipitation record for the Indian monsoon will be compared with results from a comprehensive climate model simulating the Holocene and the Eemian.

2 Materials and Methods

The model used in the present study is the Kiel Climate Model (KCM; Park et al. 2009), which is a coupled atmosphere-ocean-sea ice general circulation model. The atmosphere is represented by ECHAM5 (Roeckner et al. 2003) using the numerical resolution T31L19 corresponding to 3.75° on a great circle. ECHAM5 is coupled to the ocean model NEMO, consisting of the OPA9 ocean circulation (Madec 2008) and the LIM2 sea-ice model (Fichefet and Morales Maqueda 1997) with a horizontal resolution of approx. 2° × 2° and increased meridional resolution (0.5°) close to the equator. We measured a first set of samples for δD and δ^{13}C of sedimentary leaf wax *n*-alkanes from a marine sediment core collected on the Bengal deep-sea fan (core SO188 17286-1; 19°44′48″ N, 89°52′76″ E) as paleoindicators for precipitation and vegetation changes, respectively. Analysis of δD and δ^{13}C of *n*-alkanes is described in detail by Wang et al. (2013). A preliminary chronology for this core covering the last 135 ka was established by tuning the δ^{18}O record of planktonic foraminifera to the δ^{18}O curve of the Greenland NGRIP ice core (NGRIP-members 2004) until 126 ka BP and to SPECMAP (Martinson et al. 1987) and between 115 ka BP and 135 ka BP to fully cover the Eemian.

Simulations with the KCM (Park et al. 2009) were carried out as quasi steady state (time slice) simulations forced by changes in orbital parameters for the respective early and late Holocene and Eemian periods, corresponding to preindustrial, 9.5 ka BP and 126 ka BP (Khon et al. 2010, 2012; Schneider et al. 2010; Salau et al. 2012). Modeled precipitation is integrated over the Ganges-Brahmaputra catchment area in order to be directly comparable with data from the sediment record located in the Bay of Bengal, which is strongly influenced by this major river system. Furthermore, the terrestrial vegetation model Biome4 (Haxeltine and Prentice 1996; Kaplan et al. 2003) was forced with input from the KCM to simulate climate-induced changes in the distribution of plant functional types (PFT), whereby the C_4/C_3 balance was defined as the fraction of net primary production produced by C_4 PFT. For comparison, past changes in the relative proportions of C_4 and C_3 plants can be derived from the $\delta^{13}C$ isotopic signature of leaf wax n-alkanes measured in marine sediments. As a sensitivity test we here show that air temperature, precipitation, and atmospheric CO_2 levels are the most important climate factors influencing the C_4/C_3 signature of terrestrial vegetation, thus finally shaping the $\delta^{13}C$ signal registered in the sediment core.

3 Key Findings

The Indian summer monsoon is one of the major systems contributing to tropical climate variability. In particular, there are large seasonal variations with intensive rainfall during summer. The δD of multiple n-alkanes, a proxy for the amount of continental rainfall, implies increased precipitation during the early Holocene and the Eemian at a similar magnitude (Fig. 1a). Also Ti/Ca ratios, a proxy for litho-genic vs. marine input (not shown), support the hypothesis for enhanced monsoon rainfall since higher precipitation and river run off are the primary driver for larger terrigenous mineral fluxes during the Holocene and Eemian. The $\delta^{13}C$ of four individual n-alkanes is shifted towards more depleted values during the interglacials compared to glacial values, indicating an expansion of C_3 vegetation or/and increased humidity in a warmer climate (Fig. 1a). The differences between $\delta^{13}C$ values for individual n-alkanes in each sample analysed are very likely due to the different proportional contributions from different vegetation types (Wang et al. 2013). However, quite similar $\delta^{13}C$ values within 2 ‰ for Holocene and Eemian samples suggest similar vegetation types during both interglacials. Our preliminary results imply that despite the difference in insolation forcing, the amelioration of vegetation and increase in monsoonal precipitation over India experienced similar magnitudes for the Holocene and the Eemian.

According to climate model simulations the strength of the Indian Monsoon was more intense in both the early Holocene and the early Eemian compared to pre-industrial (Fig. 1b, lower panels). Although both interglacial periods underwent comparable temporal changes in orbital configurations, an overall higher eccen-tricity during the Eemian resulted in a greater magnitude of insolation change than during the Holocene (Fig. 1a). Consequently, the mean annual summer precipita-tion (June-August; JJA) is enhanced by up to 7 mm/day in the Early Holocene and by up to 9 mm/day during the Early Eemian relative to preindustrial in our model simulations. Also changes in net precipitation (precipitation-evaporation) indicate wetter conditions for the Eemian than for the Holocene, a general northern hemi-sphere pattern which also holds true for the African and South American Monsoon

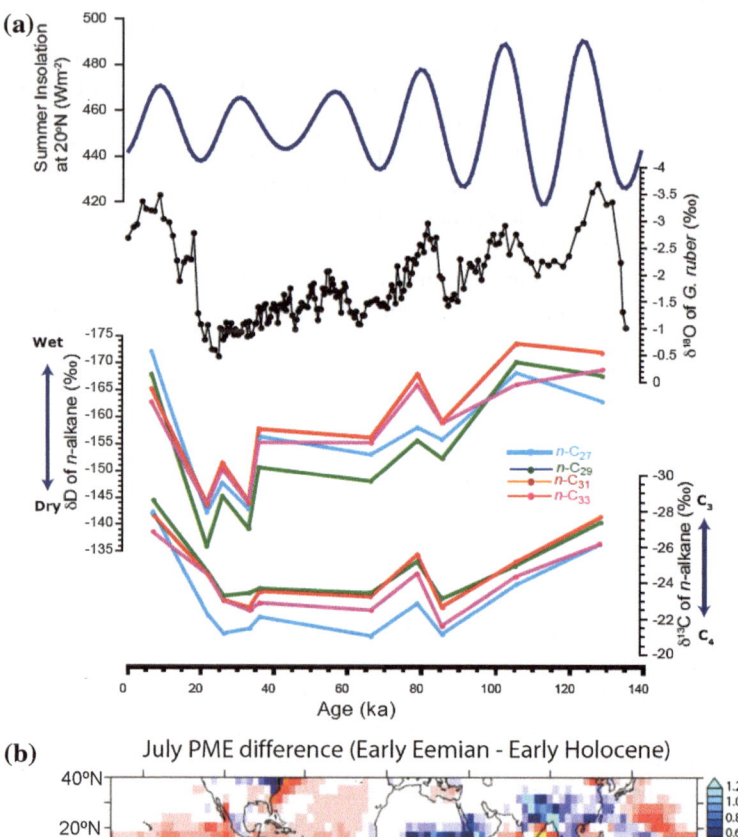

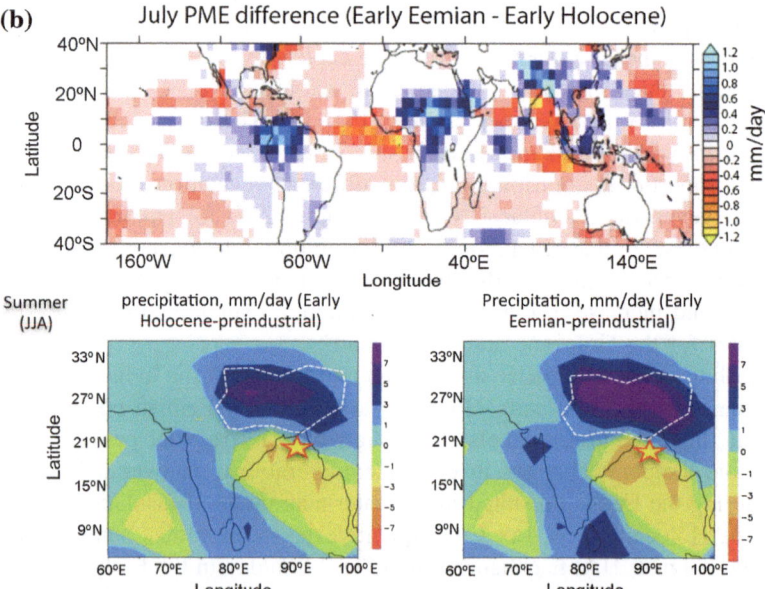

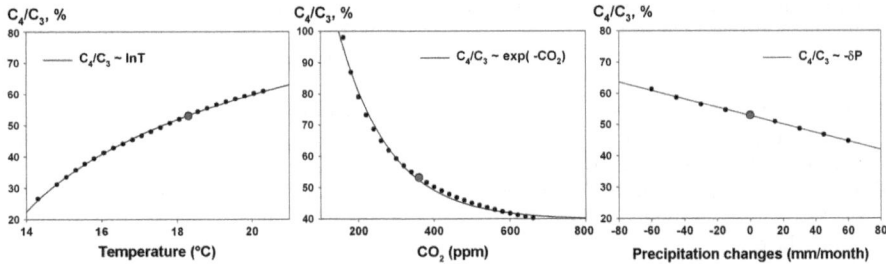

Fig. 1 a Summer (JJA) insolation at 20°N (*blue line*) (Laskar et al. 2004) and paleoclimatic records of core SO 188 17286-1 versus age for the last 135 ka: planktonic foraminiferal *G. ruber* $\delta^{18}O$ record (*black solid line*); δD and $\delta^{13}C$ records of the four most abundant *n*-alkanes. More negative values of δD indicate higher precipitation and vice versa. More negative values of $\delta^{13}C$ records indicate higher proportions of C_3 vegetation, whereas enriched values indicate increase in the contribution of C_4 vegetation. **b** Simulated difference between early Eemian and early Holocene precipitation minus evaporation balance for July (*upper panel*) and difference in total precipitation (JJA) for the early Holocene (*left bottom*) and Eemian (*right bottom*) relative to Present

Fig. 2 Sensitivity of the vegetation C_4/C_3 ratio to changes in temperature (*left*), atmospheric CO_2 (*middle*), and precipitation (*right*). *Red dots* represent the modern situation

systems (Fig. 1b, upper panel). The C_4/C_3 ratio of vegetation is determined by climate (temperature, precipitation) and atmospheric CO_2. Our model results highlight a non-linear response in the abundance of C_4 vegetation expressed as a logarithmic increase in C_4/C_3 plant proportions with rising temperatures and an exponential decrease under higher atmospheric CO_2 concentrations (Fig. 2). The response to precipitation change is rather linear with drier conditions favoring a higher C_4 vegetation ratio (Fig. 2). Any shift in vegetation C_4/C_3 ratios inferred from $\delta^{13}C$ of sedimentary *n*-alkanes is thus a combination of potentially opposing effects and cannot straightforwardly be interpreted as response to humid or dry conditions only. As a consequence, also more efforts have to be spent to regional calibration and application of the δD of individual *n*-alkanes, because it is still difficult to interpret the sedimentary δD signal even of individual *n*-alkanes from plant waxes as a quantitative measure for past changes in precipitation independent of vegetation changes (Wang et al. 2013). Obviously, combinatory studies including multiproxy records for past precipitation and river runoff, as well as coupled ocean-atmosphere-vegetation models provide the best approach to gain better insights to the complexity of and the past variations in the Indian summer monsoon system.

References

Braconnot P, Marzin C, Gregoire L, Mosquet E, Marti O (2008) Monsoon response to changes in Earth's orbital parameters: comparisons between simulations of the Eemian and of the Holocene. Clim Past 4:281–294

Cruz FW, Burns SJ, Karmann I, Sharp WD, Vuille M, Cardoso AO, Ferrari JA, Silva Dias PL, Viana O (2005) Insolation-driven changes in atmospheric circulation over the past 116,000 years in subtropical Brazil. Nature 434(7029):63–66

Cruz FW, Vuille M, Burns SJ, Wang X, Cheng H, Werner M, Edwards L, Karmann I, Auler AS, Nguyen H (2009) Orbitally driven east-west antiphasing of South American precipitation. Nat Geosci 2(3):210–214

Fichefet T, Morales Maqueda MA (1997) Sensitivity of a global sea ice model to the treatment of ice thermodynamics and dynamics. J Geophys Res 102:12609–12646. doi:10.1029/97JC00480

Fleitmann D, Burns SJ, Neff U, Mangini A, Matter A (2003) Changing moisture sources over the last 330,000 years in Northern Oman from fluid-inclusion evidence in speleothems. Quat Res 60(2):223–232

Haxeltine A, Prentice IC (1996) BIOME3: an equilibrium terrestrial biosphere model based on ecophysiological constraints, availability, and competition among plant functional types. Global Biogeochem Cycles 10:693–709

IPCC (2007) Climate change 2007: synthesis report. contribution of working groups I, II and III to the fourth assessment report of the intergovernmental panel on climate change. IPCC, Geneva, Switzerland

Kaplan JO, Bigelow NH, Prentice IC, Harrison SP, Bartlein PJ, Christensen TR, Cramer W, Matveyeva NV, McGuire AD, Murray DF, Razzhivin VY, Smith B, Walker DA, Anderson PM, Andreev AA, Brubaker LB, Edwards ME, Lozhkin AV (2003) Climate change and Arctic ecosystems: 2. Modeling, paleodata-model comparisons, and future projections. J Geophys Res. doi:10.1029/2002JD002559

Khon VC, Park W, Latif M, Mokhov I, Schneider B (2010) Response of the hydrological cycle to orbital and greenhouse gas forcing. Geophys Res Lett. doi:10.1029/2010GL044377

Khon VC, Park W, Latif M, Mokhov I, Schneider B (2012) Tropical circulation and hydrological cycle response to orbital forcing. Geophys Res Lett. doi:10.1029/2012GL052482

Kukla GJ, Bender ML, de Beaulieu JL, Bond G, Broecker WS, Cleveringa P, Gavin JE, Herbert TD, Imbrie J, Jouzel J, Keigwin LD, Knudsen KL, McManus JF, Merkt J, Muhs DR, Müller H, Poore RZ, Porter SC, Seret G, Shackleton NJ, Turner C, Tzedakis PC, Winograd IJ (2002) Last interglaical climates. Quat Res 58:2–13

Kutzbach JE, Liu X, Liu Z, Chen G (2008) Simulation of the evolutionary response of global summer monsoons to orbital forcing over the past 280,000 years. Clim Dynam 30:567–579

Laskar J, Robutel P, Joutel F, Gastineau M, Correia ACM, Levrard B (2004) A long-term numerical solution for the insolation quantities of the Earth. Astron Astrophys 428:261–285

Leduc G, Schneider R, Kim JH, Lohmann G (2010) Holocene and Eemian sea surface temperature trends as revealed by alkenone and Mg/Ca paleothermometry. Quat Sci Rev 29(7–8):989–1004

Madec G (2008) NEMO ocean engine. Notes Pole Model. 27, Institute Pierre-Simon Laplace, Paris

Martinson DG, Pisias NG, Hays JD, Imbrie JD, Moore TC, Shackleton NJ (1987) Age dating and the orbital theory of the Ice Ages: development of a high-resolution 0 to 300,000-year chronostratigraphy. Quat Res 27:1–29

NGRIP-members (2004) High-resolution record of Northern Hemisphere climate extending into the last interglacial period. Nature 431(7005):147–151

Park W, Keenlyside N, Latif M, Ströh A, Redler R, Roeckner E, Madec G (2009) Tropical pacific climate and its response to global warming in the kiel climate model. Clim Dynam doi: 10.1175/2008JCLI2261.1

Roeckner E, Bauml G, Bonaventura L, Brokopf R, Esch M, Giorgetta M, Hagemann S, Kirchner I, Kornblueh L, Manzini E, Rhodin A, Schlese U, Schulzweida U, Tompkins A (2003) The atmospheric general circulation model ECHAM 5. Part I: Model description, Report No. 349. Max Planck Institute For Meteorology, Hamburg

Salau OR, Schneider B, Park W, Khon V, Latif M (2012) Modeling the ENSO impact of orbitally induced mean state climate changes. J Geophys Res. doi:10.1029/2011JC007742

Schneider B, Leduc G, Park W (2010) Disentangling seasonal signals in Holocene climate trends by satellite-model-proxy integration. Paleoceanography. doi:10.1029/2009PA001893

Wang YJ, Cheng H, Edwards RL, An ZS, Wu JY, Shen CC, Dorale JA (2001) A high-resolution absolute-dated Late Pleistocene monsoon record from Hulu Cave. China Sci 294 (5550):2345–2348

Wang YJ, Cheng H, Edwards RL, Kong X, Shao X, Chen S, Wu JY, Jiang X, Wang X, An ZS (2008) Millennial- and orbital-scale changes in the East Asian monsoon over the past 224,000 years. Nature 451(7182):1090–1093

Wang YV, Larsen T, Leduc G, Andersen N, Blanz T, Schneider RR (2013) What does leaf wax δD from a mixed C3/C4 vegetation region tell us? Geochim Cosmochim Acta 111:128–139

Vegetation, Climate, Man—Holocene Variability in Monsoonal Central Asia

Anne Dallmeyer, Ulrike Herzschuh, Martin Claussen, Jian Ni, Yongbo Wang, Steffen Mischke and Xianyong Cao

Abstract We have investigated the Holocene climate and vegetation change in the Asian monsoon region using climate model simulations and proxy derived vegetation and climate reconstructions. The simulated mid-Holocene climate is qualitatively in good agreement with the reconstructions. Both methods reveal no systematic and uniform large-scale climate shifts, but asynchronous moisture changes in different sub-areas of the Asian monsoon region. The atmospheric response to the Holocene insolation forcing is strongly modified by ocean-atmosphere interactions, while the interaction between vegetation and atmosphere has minor influence on the large-scale Holocene climate change and is only important at a regional level. Nevertheless, sensitivity simulations reveal that large-scale forest decline in the Asian monsoon region leads to substantial losses in regional precipitation. During the Holocene, substantial vegetation changes are confined to the fringe zone of the Asian monsoon area and to the Tibetan Plateau, where simulated forest fraction has decreased by approx. 15 % and 30 % since mid-Holocene, respectively.

Keywords Holocene · Asian monsoon · Land cover change · Vegetation-atmosphere interaction · Pollen-based reconstructions · Paleo-climate modeling

A. Dallmeyer (✉) · M. Claussen
Max Planck Institute for Meteorology, Hamburg, Germany
e-mail: anne.dallmeyer@mpimet.mpg.de

U. Herzschuh · J. Ni · Y. Wang · X. Cao
Helmholtz Centre for Polar and Marine Research, Alfred Wegener Institute, Potsdam, Germany

U. Herzschuh · Y. Wang · S. Mischke · X. Cao
Institute of Earth and Environmental Science, University of Potsdam, Potsdam, Germany

M. Claussen
Meteorological Institute, University of Hamburg, Hamburg, Germany

J. Ni
Institute of Biochemistry and Biology, University of Potsdam, Potsdam, Germany

S. Mischke
Institute of Geological Sciences, Freie Universität Berlin, Berlin, Germany

© The Author(s) 2015
M. Schulz and A. Paul (eds.), *Integrated Analysis of Interglacial Climate Dynamics (INTERDYNAMIC)*, SpringerBriefs in Earth System Sciences, DOI 10.1007/978-3-319-00693-2_16

1 Introduction

Monsoon circulations are primarily driven by the strong land-sea temperature and pressure gradients that seasonally reverse following the variations in incoming solar radiation. As coupled atmosphere-ocean-land phenomena, monsoon systems are sensitive to changes in all components of the climate system and are therefore characterized by a strong temporal variability, covering multi-millennial to intra-seasonal timescales. The Asian monsoon is the most complex monsoon as it includes two sub-monsoon systems, i.e., the Indian and the East Asian monsoon, that partly interact with one another, partly counteract each other and react differently to internal forcings entailing spatial climate variability within the Asian monsoon region. Furthermore, the Asian monsoon domain includes the Tibetan Plateau which affects the regional and global circulation, as well as the water and energy cycle due to its large horizontal and vertical extent.

Within this project, we investigated the climate and vegetation variability in the Asian monsoon region during the Holocene using reconstructions and climate model simulations. Special focus was set on the investigation of the role of feedbacks in the Holocene climate change, particularly the role of the interaction between vegetation and atmosphere. Our results suggest that humans have only become an effective driver of large-scale vegetation change within the last 1,000 years. Therefore this topic had been given minor attention in our research and will not be discussed in this chapter.

2 Materials and Methods

The modeling results of our project are based on different simulations performed in the global atmosphere-ocean-vegetation model ECHAM5/JSBACH-MPIOM. To investigate the Holocene vegetation and climate trend, we analyzed the transient simulation, covering the last 6,000 years, undertaken by Fischer and Jungclaus (2011, cf. Chap. Evaluation of Eemian and Holocene Climate Trends: Combining Marine Archives with Climate Modelling). The climate data of this and other transient simulations has additionally been used as forcing for the diagnostic biome model BIOME4. The analysis of the contribution of ocean-atmosphere and vegetation-atmosphere interactions to the Holocene climate change was conducted using a set of simulations designed for factor separation by Otto et al. (2009). This set was extended by sensitivity simulations with fixed ocean and different idealized land cover prescribed (Dallmeyer and Claussen 2011). In addition, we performed high resolution (T106L31) time slice experiments for the early Holocene (9k), mid-Holocene (6k) and pre-industrial (0k) climate with prescribed sea-surface temperatures (SSTs) and sea-ice (Dallmeyer et al. 2013).

As part of this project pollen records from all over continental eastern Asia were collected and taxonomically harmonized (Cao et al. 2013). Age-depth models were re-established using the "Bacon" software in R and selecting the IntCal09 radiocarbon calibration curve for radiocarbon dating data calibrating. The interpolated pollen percentages for individual taxa were obtained for each 500-year interval between 22 cal. thousand years (ka) before present (BP) and 0 cal. ka BP using the linear integration function of AnalySeries 2.0.4.2. The distributions and qualities of these available pollen records were presented using ArcMap 10. Paleoclimate records covering the past 18 cal. ka BP were collected from monsoonal Central Asia (Wang et al. 2010). After recalibrating radiocarbon ages, semi-quantitative climate indices for moisture and temperature were calculated in 100-year intervals for the Holocene and 200-year intervals for the late glacial period. Principal component analysis was applied using the moisture index covering the last 10 cal. ka BP to examine the climate patterns.

3 Key Findings

Climate change: Reconstructions reveal strong spatial heterogeneity in Holocene moisture evolution within the Asian monsoon region. In the Indian monsoon domain, moisture levels are highest during the early Holocene while many records from the East Asian monsoon region (particularly north-central China) do not reach maximum wetness until mid-Holocene (Wang et al. 2010). For the early and mid-Holocene time-slices, the high-resolution climate model results are qualitatively in good agreement with the reconstructions (6k: Dallmeyer et al. 2013): Both show wetter conditions in the Indian monsoon region and a spatially inhomogeneous pattern of precipitation change in the East Asian monsoon region (Fig. 1).

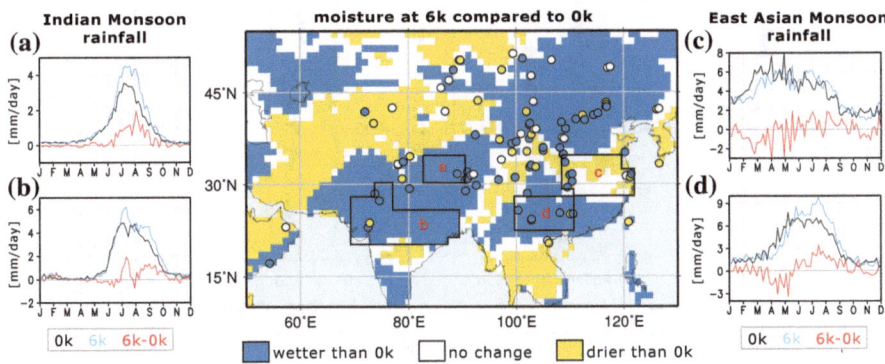

Fig. 1 Map: Simulated (*shaded*) and reconstructed (*dots*) mid-Holocene (6k) moisture difference to pre-industrial level (0k) adapted to a 3-scale moisture index that categorize the mid-Holocene climate in 'wetter than 0k', 'drier than 0k' and 'similar as 0k', i.e., no change (absolute value of relative precipitation difference less than 3 % in the model). Small figures **a–d**: climatological annual precipitation cycle (mm/day) at mid-Holocene (*blue*), pre-industrial (*black*) and mid-Holocene minus pre-industrial (*red*) for different areas in the Indian (**a**, **b**) and East Asian monsoon region (**c**, **d**). Please note the change in scales. For further details see Dallmeyer et al. (2013)

According to the model, these dissimilarities result from the different nature of the two sub-monsoon systems. In the Indian monsoon area, most of the annual precipitation and, hence, of the precipitation change is related to the summer monsoon. In the East Asian monsoon domain, it rains during summer and spring, so that the sign of the annual signal is determined by the balance of decreased spring and increased summer precipitation during mid-Holocene (Fig. 1). This ratio could be strongly affected by the local environment (e.g. topography). Our study shows, that moisture changes in the East Asian monsoon system can not be interpreted as an indicator for summer monsoon intensity changes.

Vegetation change: According to vegetation modeling and the fossil data set including 218 harmonized and homogenized pollen records (Cao et al. 2013), the Holocene vegetation change in the Asian monsoon region is rather small. Large changes in land cover only occur in the transition zone of the moist Asian summer monsoon and the dry westerly winds in East Asia as well as on the Tibetan Plateau. For the Tibetan Plateau, high resolution pollen records (presented and discussed in Herzschuh et al. 2010) are compared to the transient vegetation simulation (Dallmeyer et al. 2011). The simulated and reconstructed vegetation trends agree for most sites and both reveal a strong vegetation degradation and forest decline on the Tibetan Plateau (Fig. 2). The simulated forest fraction decreases by nearly one third from mid-Holocene to pre-industrial, revealing that the land cover on the

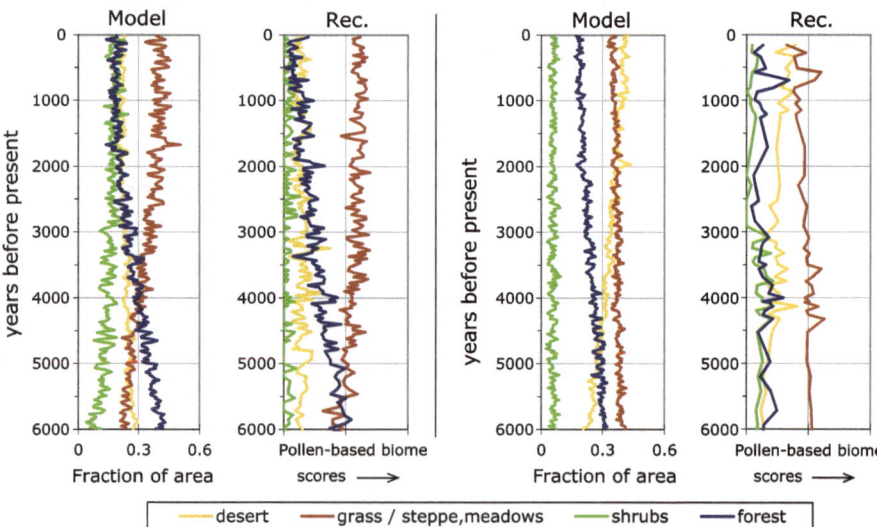

Fig. 2 Simulated (Model) land cover trend (fraction of area, 1 = 100 %) and reconstructed (Rec.) land cover trend (arbitrary units) from mid-Holocene (6,000 years BP) to pre-industrial (0k) for two of the four analyzed regions on the Tibetan Plateau (for details see: Dallmeyer et al. 2011). The biome reconstructions are given in arbitrary units called biome scores and they only give semi-quantitative information on dominant and sub-dominant biomes. *Left panel*: North-Eastern Tibetan Plateau compared to Lake Qinghai (36.55° N, 100.1° E). *Right panel*: Central-Western Tibetan Plateau compared to Lake Bangong (33.42° N, 79° E)

Tibetan Plateau is very sensitive to climate change. The model results reveal that besides precipitation also temperature has to be considered as a driving mechanism for vegetation in monsoon-influenced regions.

On average, the biome simulations show a 15 % reduction of forest and a 20 % increase of desert in the monsoon-westerly wind transition zone from mid-Holocene to pre-industrial, but the amplitude of this signal strongly depends on the prescribed climate forcing. The desert-steppe margin is located further westward by approx. 5° at 6k, ranging from 1° to 10° in the different simulations. The forest biomes extend further north-westward by approx. 2°, ranging from 0° to 4°.

Role of vegetation-atmosphere interaction: Most of the temperature and precipitation changes in the Asian monsoon region can be attributed to the direct response of the atmosphere to the Holocene insolation forcing. However, this direct signal is strongly modified by the ocean-atmosphere interaction (Dallmeyer et al. 2010). In India, the interactive ocean strongly increases the Indian summer monsoon precipitation (0.68 mm/day) and can even overcompensate the decreased precipitation calculated in the atmosphere-only run (−0.22 mm/day) for the mid-Holocene. The East Asian monsoon weakens due to the atmosphere-ocean interaction, particularly above the ocean. The vegetation-atmosphere interaction generally has a minor effect on the Holocene climate change in our simulations. However, the idealized sensitivity simulations reveal that large-scale forest decline in East and South Asia leads to heavy losses in precipitation at a regional level (Dallmeyer and Claussen 2011). For instance, in the area along the yellow river, which was one of the major settlement areas of prehistoric cultures in Asia, the simulated precipitation change related to forest decrease is half as large as the insolation induced signal and may therefore have strongly contributed to the attenuating Asian summer monsoon precipitation known from reconstructions.

References

Cao X, Ni J, Herzschuh U, Wang Y, Zhao Y (2013) A late Quaternary pollen dataset from eastern continental Asia for vegetation and climate reconstructions: set up and evaluation. Rev Palaeobot Palyno 194:21–37

Dallmeyer A, Claussen M (2011) The influence of land cover change in the Asian monsoon region on present-day and mid-Holocene climate. Biogeosciences 8:1499–1519

Dallmeyer A, Claussen M, Otto J (2010) Contribution of oceanic and vegetation feedbacks to Holocene climate change in monsoonal Asia. Clim Past 6:195–218

Dallmeyer A, Claussen M, Herzschuh U, Fischer N (2011) Holocene vegetation and biomass changes on the Tibetan Plateau - a model-pollen data comparison. Clim Past 7:881–901

Dallmeyer A, Claussen M, Wang Y, Herzschuh U (2013) Spatial variability of Holocene changes in the annual precipitation pattern: a model-data synthesis for the Asian monsoon region. Clim Dyn 40(11):2919–2936

Fischer N, Jungclaus JH (2011) Evolution of the seasonal temperature cycle in a transient Holocene simulation: orbital forcing and sea-ice. Clim Past 7:1139–1148

Herzschuh U, Birks HJB, Ni J, Zhao Y, Liu H, Liu X, Grosse G (2010) Holocene land-cover changes on the Tibetan Plateau. Holocene 20:91–104

Otto J, Raddatz T, Claussen M (2009) Climate variability-induced uncertainty in mid-Holocene atmosphere-ocean-vegetation feedbacks. Geophys Res Lett 36:L23710. doi:10.1029/2009GL041457

Wang YB, Liu XQ, Herzschuh U (2010) Asynchronous evolution of the Indian and East Asian summer monsoon indicated by Holocene moisture patterns in monsoonal central Asia. Earth Sci Rev 103(3–4):135–153. doi:10.1016/j.earscirev.2010.09.004

Global Land Use and Technological Evolution Simulations to Quantify Interactions Between Climate and Pre-industrial Cultures

Carsten Lemmen, Kerstin Haberkorn, Richard Blender,
Klaus Fraedrich and Kai W. Wirtz

Abstract To understand the two-way interaction between past societies and Holocene climate, we conduct a series of integrated model- and data-based studies. The climate-culture feedback is investigated using a coupled Earth System Civilization Model, including a new methodology to incorporate proxy information into an Earth System Model. Our study reconstructs the transition to agriculture for Western Eurasia in the paleoclimatic context; it shows that migration is not a necessary prerequisite for this transition, which is a yet unresolved problem in European archeology. Climate variability and extreme events had no significant impact, which reflects societal resilience. Also, our simulation studies indicate a considerable range of global and regional carbon emissions by deforestation. In conclusion, we find on the one hand a lower sensitivity of past societies to changes in Holocene climate than frequently suggested, on the other hand a possibly larger influence of those societies on regional and global climate.

Keywords Neolithic transition · Preindustrial cultures · Earth system model · Sociotechnological model · Anthropocene · Proxy integration · Adaptation · Carbon emission

1 Introduction

Key questions of Holocene climate and its interaction with pre-industrial cultures are: How much did climate variability determine where and when agriculture appeared or cultures disappeared? When did humans start to interfere with and how

C. Lemmen (✉) · K.W. Wirtz
Institute for Coastal Research, Helmholtz-Zentrum Geesthacht, Geesthacht, Germany
e-mail: carsten.lemmen@hzg.de

C. Lemmen
C.L. Science Consult, Lüneburg, Germany

K. Haberkorn · R. Blender · K. Fraedrich
Meteorological Institute, University of Hamburg, Hamburg, Germany

© The Author(s) 2015
M. Schulz and A. Paul (eds.), *Integrated Analysis of Interglacial Climate Dynamics (INTERDYNAMIC)*, SpringerBriefs in Earth System Sciences,
DOI 10.1007/978-3-319-00693-2_17

much did they disturb global and regional carbon and hydrological cycles? When did the anthropocene begin (Kaplan et al. 2011)? To understand the two-way interaction between past societies and Holocene climate defines a challenge for transdisciplinary research and for testing controversial hypotheses of a middle Holocene influence of humans on climate.

We addressed these questions by using an interactively coupled model system composed of a cultural adaptation model [Global Land Use and technological Evolution Simulator, GLUES (Lemmen and Wirtz 2010, 2012; Lemmen et al. 2011)] and an Earth System Model [Planet Simulator, PLASIM (Haberkorn 2013; Haberkorn et al. 2012)]. Cultural feedback on climate is implemented by land surface changes. The realism of the interactive simulation of climate and culture is improved by constraining the climate model with temperature and precipitation proxies, and by constraining the cultural model with archeological data compilations and vegetation proxies indicative of human land use. Abrupt climate changes are included based on globally available time series of proxy-derived climate variability (Wirtz et al. 2010). Models and data employed in our study cover the period 11.5 thousand years (ka) before present (BP) to 3 ka BP and are global in scope. Carbon emissions are evaluated regionally, and the sociotechnological model is validated against regional archeological data for Northern Central Europe and South Asia.

We propose to integrate the dynamic anthroposphere into today's state-of-the-art Earth System Models (ESM) as a prerequisite to better understand current human-climate interaction and adaptation to ongoing climate change. Current and anticipated users of our work are paleoclimate and paleovegetation modelers, paleoclimate variability analysts, archeologists, and agricultural economists.

2 Materials and Methods

A large data set of 235 long-term (>4,000 years) and high-resolution (mostly <100 years) time series of climate information have been collected from the literature and Interdynamic partners. Based on a change-point analysis, we partitioned the Holocene into slightly overlapping periods, the early Holocene (11–5 ka BP) and late Holocene (6–0 ka BP). For each interval, we evaluated each proxy time series for statistically significant periodic signals, using very strict and data-adaptive thresholds for significance.

For simulations, we chose the PLASIM (Haberkorn 2013; Haberkorn et al. 2012) ESM which can be used to run climate simulations for multi-millennial time scales in acceptable real time while relying on a fully dynamic core; it also offers different vegetation couplers (Haberkorn 2013 and Fig. 2). We performed full Holocene transient simulations at T21 and T42 resolutions with orbital, greenhouse gas (GHG) forcing, and climatological sea-surface temperature (SST). A novel scheme was devised to reconstruct past SST from the sensitivity of land temperature to SST diagnosed from a comparison between present day climate and present day proxy climate. Socio-ecological adaptations to climate change are modeled with GLUES (Lemmen et al. 2011; Wirtz and Lemmen 2003); and the land-use feedback on population is simulated by overexploitation of land and resources. Characteristic traits of technology, substance, and economic potential exhibit adaptation and continuous innovation (Lemmen 2014).

Paleovegetation and paleoclimate forcings of GLUES, expressed as net primary production and growing degree days, are derived from PLASIM. After successfully testing vegetation fields (Haberkorn 2013), we ran GLUES to obtain socio-technological trajectories for regional subdomains, worldwide and over the entire Holocene.

3 Key Findings

The average size of regionally coherent climate variability derived from paleoclimate proxies is around 3,000 km; it significantly increased over Western South and North America, and decreased over the Arabian Sea and the Southwest Asian monsoon region. Centennial (but not millennial) scale variability decreased over the North Atlantic: this reinforced our earlier hypothesis (Wirtz and Lemmen 2003) that regional climate variability may have led to unequal probabilities for crises in early human civilizations in the Old and the New World.

The high-resolution vegetation distribution during the 6 ka BP time slice, based on the PLASIM climate, reveals generally deciduous and temperate vegetation types in Western, South and Central Europe (Fig. 1). Such a high-resolution vegetation reconstruction can improve the prediction of suitability of the land for past agricultural activities (Kaplan et al. 2011).

We successfully reconstructed a North Atlantic SST that—taken as a boundary condition for PLASIM—replicates the terrestrial temperature (i.e., the temperature relevant to societies) in Central Europe as represented by a high-resolution lake proxy. This novel method applies an inverse modeling algorithm to nudge the simulated land temperature in the climate model to a proxy-reconstructed temperature; its central element is the inverse climate sensitivity in Central Europe to North Atlantic SST anomalies (Haberkorn 2013; Haberkorn et al. 2012). We can now provide past SST fields that are in agreement with reconstructed land temperature, thus allowing the reconstruction of a dynamically consistent European climate throughout the Holocene.

We calculated the prehistoric GHG emissions from anthropogenic land use; we produced estimates for land demand for crops (Gaillard et al. 2010) and associated carbon emissions (Lemmen 2010) for the Holocene. Calculated emissions (world total 30 Gt by 4 ka BP from deforestation) could not have contributed to a significant warming. By considering past technological inefficiencies, however, we arrived at much larger emissions on the order of 340 Gt by 100 years BP (AD 1850) (Kaplan et al. 2011), consistent with the stable carbon isotope signature from ice cores. Thus, our two studies provide extreme low and high estimates of the possible

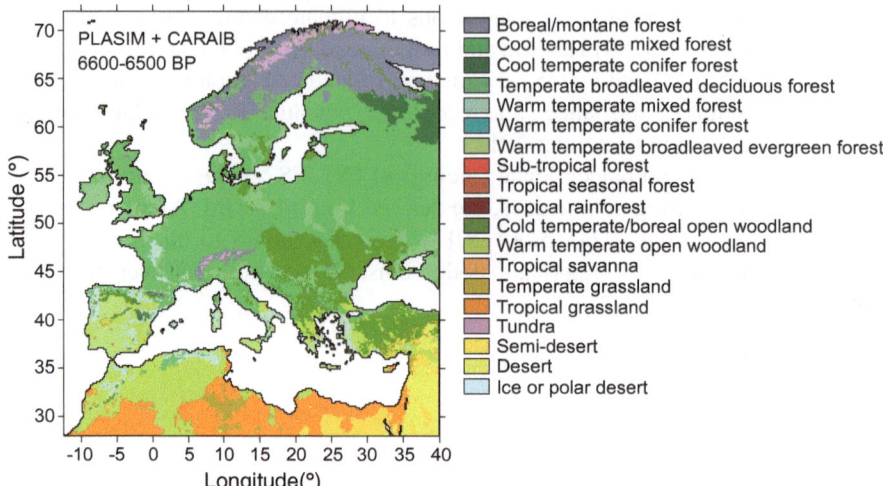

Fig. 1 European vegetation distribution for 6 ka BP, simulated by the high-resolution vegetation model CARAIB forced with PLASIM climate

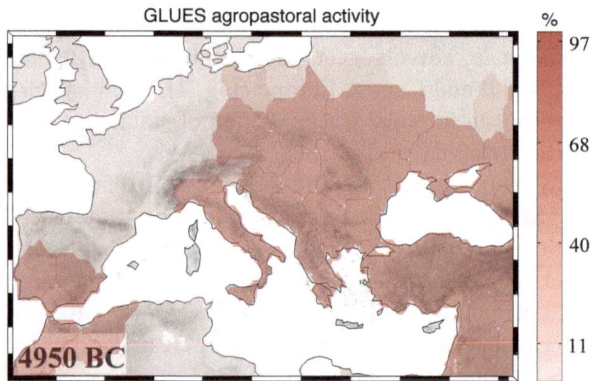

Fig. 2 Fraction of agricultural activity versus foraging subsistence at a critical transition time (6,900 years BP/4950 BC), when the GLUES-simulated frontier between farmers and foragers ran across central Europe

range of anthropogenic activities, and the Early Anthropocene hypothesis (Kaplan et al. 2011; Lemmen 2010).

Simulations of the transition to agriculture (Fig. 2) agree with archeological site data across Western Eurasia within a model uncertainty of ± 500 years (Lemmen and Wirtz 2012; Lemmen et al. 2011). Thus, GLUES is able to realistically simulate the onset of agriculture not only on a global scale as previously reported (Wirtz and

Lemmen 2003) but also within the broader region of Europe and the Mediterranean. For the first time, a numerical model shows that the transition to agriculture can also be explained by information exchange, rather than migration (Lemmen et al. 2011).

Climate events may not have been as important for early sociocultural dynamics as endogenous factors. This could be demonstrated by using idealized climate events (Wirtz et al. 2010) to disturb societies in GLUES. Time-series anomalies were spatially weighted to assess the regional impact of abrupt climate excursions (Lemmen and Wirtz 2012); climate induced population decline can lead to loss of knowledge, and could impact regional technological development. Typical observed lags between cultural complexes were simulated only by a simulation which included climate extremes (Lemmen and Wirtz 2012; Lemmen et al. 2011).

The reason for the vulnerability of several societies to climate changes is their cultural specialization, that is, the restriction to only few different subsistence economies. Continuous maintenance of a diverse pool of technologies played an important role in determining the resilience of Neolithic populations to changing climates. In conclusion, past cultural and sociotechnological changes appear much less determined by Holocene climate variability than often suggested in the literature, while the influence of past agriculture on the global carbon cycle may have been larger than previously thought.

References

Gaillard MJ, Sugita S, Mazier F, Trondman AK, Broström A, Hickler T, Kaplan JO, Kjellström E, Kokfelt U, Kuneš P, Lemmen C, Miller P, Olofsson J, Poska A, Rundgren M, Smith B, Strandberg G, Fyfe R, Nielsen AB, Alenius T, Balakauskas L, Barnekow L, Birks HJB, Bjune A, Björkman L, Giesecke T, Hjelle K, Kalnina L, Kangur M, van der Knaap WO, Koff T, Lagerås P, Latałowa M, Leydet M, Lechterbeck J, Lindbladh M, Odgaard B, Peglar S, Segerström U, von Stedingk H, Seppä H (2010) Holocene land-cover reconstructions for studies on land cover-climate feedbacks. Clim Past 6:483–499

Haberkorn K, Lemmen C, Blender R, Fraedrich K (2012) Iterative land proxy based reconstruction of SST for the simulation of terrestrial Holocene climate. Earth Sys Dyn Disc 3:149–200

Haberkorn K (2013) Reconstruction of the Holocene climate using an atmosphere-ocean-biosphere model and proxy data. Dissertation, University of Hamburg. (urn:nbn:de:gbv:18-62566)

Kaplan JO, Krumhardt KM, Ellis EC, Ruddiman WF, Lemmen C, Goldewijk KK (2011) Holocene carbon emissions as a result of anthropogenic land cover change. Holocene 21:775–791

Lemmen C (2010) World distribution of land cover changes during pre- and protohistoric times and estimation of induced carbon releases. Géomorphol 4:303–312

Lemmen C (2014) Malthusian assumptions, Boserupian response in transition to agriculture models. In: Fischer-Kowalski M, Reenberg A, Schaffartzik A, Mayer A (eds) Ester Boserup's legacy on sustainability: orientations for contemporary research. Springer, Vienna, pp 87–97

Lemmen C, Wirtz KW (2010) Socio-technological revolutions and migration waves: re-examining early world history with a numerical model. In: Gronenborn D, Petrasch J (eds) The spread of the Neolithic to Central Europe. RGZM, Mainz, pp 25–37

Lemmen C, Wirtz KW (2012) On the sensitivity of the simulated European Neolithic transition to climate extremes. J Archeol Sci. doi:10.1016/j.jas.2012.10.023

Lemmen C, Gronenborn D, Wirtz KW (2011) A simulation of the Neolithic transition in Western Eurasia. J Archeol Sci 38:3459–3470

Wirtz KW, Lemmen C (2003) A global dynamic model for the Neolithic transition. Clim Change 59:333–367

Wirtz KW, Lohmann G, Bernhardt K, Lemmen C (2010) Mid-Holocene regional reorganization of climate variability: analyses of proxy data in the frequency domain. Paleogeogr Paleoclimatol Paleoecol 298:189–200

North-West African Hydrologic Changes in the Holocene: A Combined Isotopic Data and Model Approach

Enno Schefuß, Martin Werner, Britta Beckmann, Barbara Haese and Gerrit Lohmann

Abstract To achieve a better understanding of the hydrologic evolution of the North-West (NW) African monsoon system during the Holocene, in particular during inferred abrupt climate changes at the end of the African Humid Period (AHP), we investigated terrigenous plant lipids deposited in marine sediments offshore NW Africa. Changes in rainfall amount were estimated by compound-specific hydrogen isotope (δD) analyses. The spatial gradient of rainfall isotopic compositions is reflected in marine surface sediments. δD changes in plant waxes covering the last 100 years confirm the observed decrease in rainfall during the late twentieth century Sahel drought, and thus can be used for a quantitative calibration of δD and precipitation. δD changes in sedimentary plant waxes show no abrupt change at the end of the AHP suggesting a gradual precipitation decline. These results are supported by Holocene climate simulations using a coupled atmosphere-land surface model, which includes an explicit modeling of isotopic fractionation within the hydrological cycle.

Keywords North-West africa · Hydrology · Compound-specific deuterium analysis · Climate simulations · Isotope modeling

1 Introduction

A profound scientific debate persists on the response of the North-West (NW) African monsoon system (Fig. 1c, d) to long-term (orbital) changes at the end of the African Humid Period [AHP, 14,800–5,500 years before present (BP)]. One hypothesis favors an abrupt degradation of the vegetation in the once 'green Sahara'

E. Schefuß (✉) · M. Werner · B. Beckmann · G. Lohmann
MARUM—Center for Marine Environmental Sciences and Faculty of Geosciences,
University of Bremen, Bremen, Germany
e-mail: schefuss@uni-bremen.de

M. Werner · B. Haese · G. Lohmann
Alfred Wegener Institute, Helmholtz Centre for Polar and Marine Research,
Bremerhaven, Germany

© The Author(s) 2015
M. Schulz and A. Paul (eds.), *Integrated Analysis of Interglacial Climate Dynamics (INTERDYNAMIC)*, SpringerBriefs in Earth System Sciences, DOI 10.1007/978-3-319-00693-2_18

109

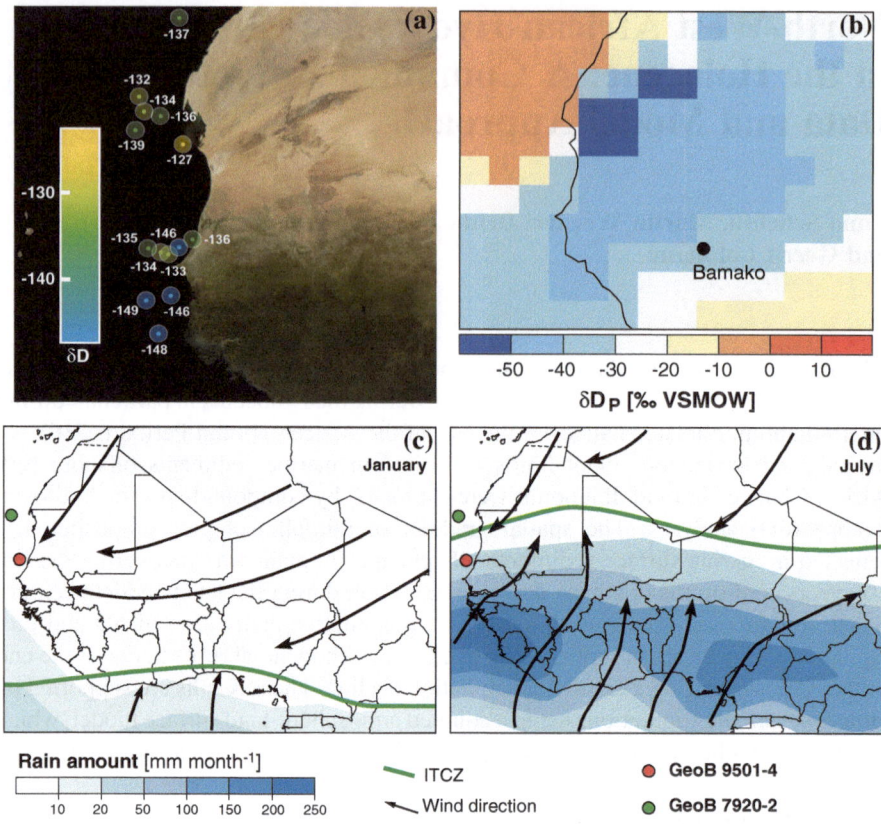

Fig. 1 a Satellite image with position of surface sediments and respective compound-specific δD composition of *n*-C$_{31}$ alkanes (*image*: NASA Earth Observations), **b** δD composition of precipitation as simulated by ECHAM5-wiso T63 simulation (nudged mode, simulated period 1959–1982), **c** Average NW African climate in January; *black arrows* display major wind directions, *green line* denotes position of the Intertropical Convergence Zone (ITCZ), *red dot* marks the position of GeoB9501-4, *green dot* marks the position of GeoB7920-2, **d** same as **c** in July

by a strong bio-geophysical climate-vegetation feedback. The slow decline in orbital-paced North African summer insolation during the Holocene might have led to a sudden, i.e., within centuries, desertification of the Sahara in the mid-Holocene (e.g., Claussen 2009). This hypothesis is supported by a rapid increase of dust export at around 5,500 years BP detected offshore NW Africa (deMenocal et al. 2000). A contrasting hypothesis suggests a continuous southward retreat of monsoonal rainfall due to insolation decline causing gradual environmental deterioration (e.g., Kröpelin et al. 2008).

To provide more insights into the paleo-hydrologic evolution of NW Africa over the Holocene and unravel past climate and vegetation changes, we conducted a combined data-modeling study. δD compositions of terrigenous plant lipids, providing direct insights into continental hydrologic changes (Sachse et al. 2012), were investigated in a marine sediment core covering the AHP in high-temporal resolution. The results were compared to the simulated rainfall amounts and its isotopic composition for selected Holocene time-slices. The simulations were performed using an atmospheric general circulation model containing explicit stable water isotope diagnostics coupled to a dynamic vegetation model. These observations and simulations provide insights into amplitudes of past climate variations and allow a quantitative assessment of precipitation changes in NW Africa during the Holocene.

2 Materials and Methods

Marine surface sediments from coastal areas around NW Africa (between 31°N to 8°N and 10°W to 20°W, Fig. 1a) were investigated to assess how the isotopic gradient in rainfall on the continent is reflected in δD compositions of terrigenous plant lipids deposited in the sediments. To understand how rainfall variations on short time-scales, i.e., during the late twentieth century Sahel drought, are reflected in δD compositions of plant-derived n-alkanes we analyse multi-core GeoB9501-4, taken near the Senegal River mouth and covering the past 100 years (Mulitza et al. 2010; Fig. 1c, d). These data allow a quantitative calibration of compound-specific δD changes against meteorological data. Sediment core GeoB7920-2 (Tjallingii et al. 2008; Fig. 1c, d), taken at the same location as ODP 658C (20°45′N, 18°35′ W; deMenocal et al. 2000), was analyzed by compound-specific δD and δ^{13}C analyses to investigate the termination phase of the AHP and reconstruct precipitation and vegetation changes. All sediment samples were extracted with organic solvents, and fractions containing plant-wax lipids were quantified by gas-chromatography and measured for their compound-specific stable carbon and hydrogen isotope compositions (details of the methods are in Schefuß et al. 2011).

For isotope modeling, stable isotopes $H_2^{18}O$ and HDO have been incorporated in the hydrological cycle of the coupled atmosphere-land surface model ECHAM5-JSBACH (Haese et al. 2013). With this newly developed setup it is possible to distinguish between evaporation and transpiration fluxes, and separately simulate the relevant fractionation processes for both. To investigate the hydrological evolution of NW Africa during the last 50 years a simulation with prescribed observed sea-surface temperature and atmospheric conditions nudged to ERA-40 reanalyses data was performed. For the Holocene, four different time-slice experiments (pre-industrial (PI), 2,000 years BP, 4,000 years BP, 6,000 years BP) were conducted and analyzed (Haese 2014).

3 Key Findings

n-Alkane distributions and concentrations in marine surface sediments from the coast of NW Africa indicate the input of terrestrial plant waxes at all locations (Fig. 1a). δD compositions of plant waxes in the sediments show lowest values offshore the Sahel and highest offshore the Sahara desert (Fig. 1a) in accordance with the latitudinal trend of δD in precipitation (Bowen and Revenaugh 2003). The approximately 25 ‰ amplitude in δD of plant waxes resembles the isotopic range of continental precipitation. The observed spatial isotopic gradient is also found in pre-industrial climate simulations with the ECHAM5-JSBACH model (Fig. 1b).

δD values of plant waxes extracted from multi-core GeoB9501-4 near the Senegal River mouth reveal temporal variations linked to hydrologic conditions over the western Sahel during the last century. The Sahel drought in the late-1960s to mid-1990s can be traced in a+10 ‰ shift of the δD signal. Changes in δD closely correspond to a rainfall index for the western Sahel (Fink et al. 2010). Comparison with precipitation-isotope data provides an estimate of a 30 % reduction in rainfall during the Sahel drought, in accordance with meteorological observations. In agreement with this data, a maximum isotopic shift of +15 ‰ in δD in precipitation is detected in the nudged ECHAM5-JSBACH simulation between wet and dry years of the period 1957–2011. However, the simulated timing of the Sahel drought is erroneous (Haese 2014).

To study the ending of the AHP, we analyze core GeoB7920-2. Terrestrial versus marine elemental ratios of GeoB7920-2 confirm an abrupt increase (within centuries) of terrigenous material around 5,500 years BP (Fig. 2a), in accordance with the findings by deMenocal et al. (2000). Sedimentary concentrations of plant waxes increase after the AHP as expected from increased eolian transport (Fig. 2b). Compound-specific δD compositions of plant waxes (Fig. 2d) reveal a gradual change following the declining summer insolation (Fig. 2f). Plant-wax δD during the AHP are about 20 ‰ depleted relative to the latest Holocene. Using the observed modern local precipitation-isotope relation, these data suggest a doubling of rainfall during the AHP relative to today. Compound-specific $\delta^{13}C$ data indicate higher C_4 plant coverage during the AHP (Fig. 2c) and also reveal a gradual vegetation change. The latter is, however, not contemporaneously with the pre-cipitation change arguing against strong bio-geophysical feedback mechanisms.

These findings are supported by the ECHAM5-JSBACH isotope time-slice experiments as well as a 6,000-year long (6,000 years BP to 0 years BP) transient simulation with a comprehensive climate model (Fischer and Jungclaus 2011). The latter experiment also reveals a gradual decrease of precipitation from approx. 5,500 years BP until 1,500 years BP in the area between 5°N and 30°N (Fig. 2e). Due to the simulated evolution of precipitation the modeled vegetation cover in NW Africa is reduced, leading to a gradual increase of the desert fraction. The four isotope time-slice ECHAM5-JSBACH experiments, forced with oceanic and veg-etation boundary conditions derived from the transient simulation by Fischer and Jungclaus (2011), indicate a change of δD in precipitation of about 15 ‰ between 6,000 years BP and pre-industrial. These modelled values agree well with the findings from the measured compound-specific δD compositions of plant waxes (Fig. 2d).

In consequence, we infer from this new data-model study that continental rainfall diminished gradually over NW Africa during the Holocene leading to progressive aridification. We find no parallel changes in vegetation type, arguing

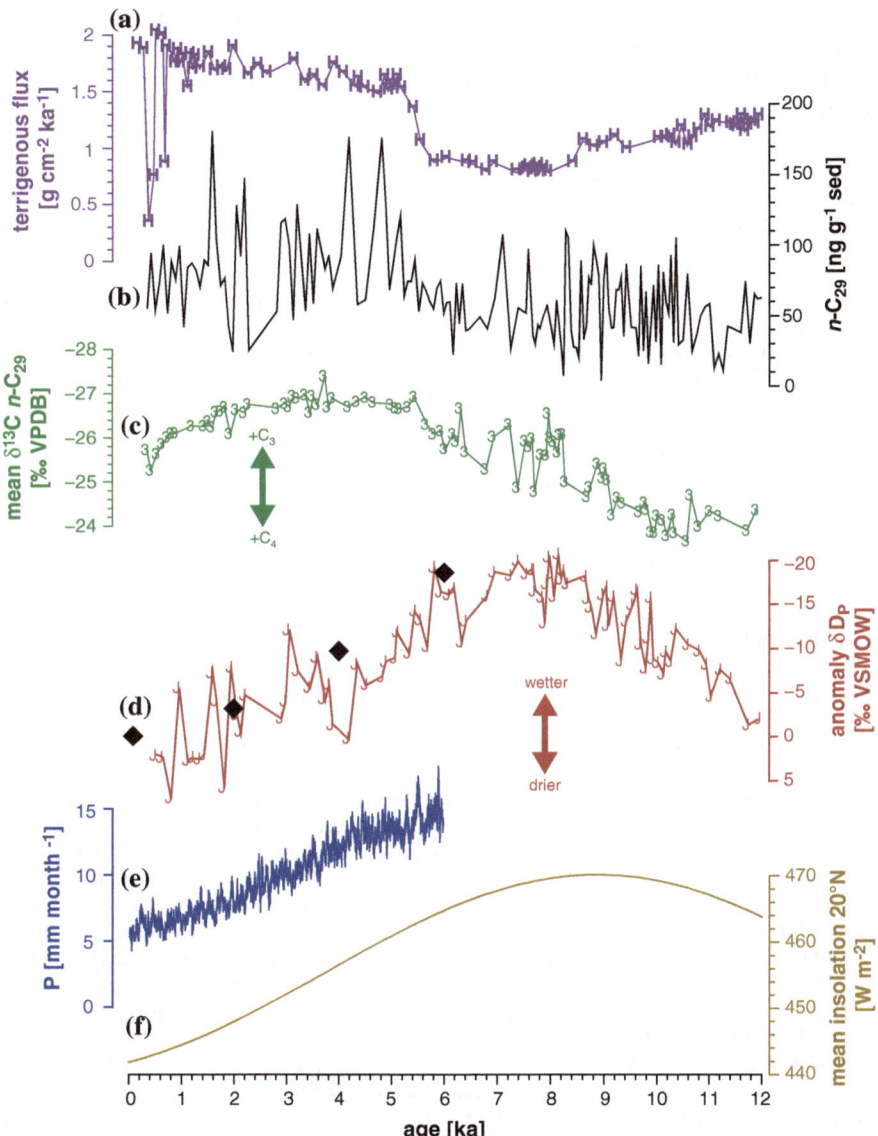

Fig. 2 **a** Terrigenous flux at ODP658 showing the inferred abrupt end of the AHP (deMenocal et al. 2000), **b** Plant-wax concentration (as n-C$_{29}$ alkane) in GeoB7920-2 indicating predominant eolian transport, **c** Compound-specific δ^{13}C compositions of n-C$_{29}$ alkane showing higher C$_4$ plant contributions during the AHP, **d** Estimated anomaly of precipitation δD versus Latest Holocene indicating a gradual continental hydrologic change; diamonds show simulated mean δD$_P$ anomalies of the ECHAM5-JSBACH time slice experiments at PI, 2,000, 4,000, and 6,000 years BP, **e** Modeled NW African precipitation amount of the transient experiment between 6,000 years BP and pre-industrial by Fischer and Jungclaus (2011) shown as 30-year running mean, **f** Mean boreal summer insolation at 20° N (Laskar et al. 2004)

against strong bio-geophysical feedback mechanisms at the end of the AHP. The abrupt increase in dust flux is inferred to be an environmental response when crossing a precipitation threshold, such as a sudden activation of dust export from desiccated lake beds.

References

Bowen GJ, Revenaugh J (2003) Interpolating the isotopic composition of modern meteoric precipitation. Water Resour Res. doi:10.1029/2003WR002086

Claussen M (2009) Late quaternary vegetation-climate feedbacks. Clim Past 5(2):203–216

de Menocal P, Ortiz J, Guilderson T, Adkins J, Sarnthein M, Baker L, Yarusinsky M (2000) Abrupt onset and termination of the African Humid Period: rapid climate responses to gradual insolation forcing. Quat Sci Rev 19:347–361

Fink AH, Schrage JM, Kotthaus S (2010) On the potential causes of the nonstationary correlations between West African precipitation and Atlantic Hurricane activity. J Clim 23:5437–5456

Fischer N, Jungclaus JH (2011) Evolution of the seasonal temperature cycle in a transient Holocene simulation: orbital forcing and sea-ice. Clim Past 7:1139–1148

Haese B, Werner M, Lohmann G (2013) Stable water isotopes in the coupled atmosphere–land surface model ECHAM5-JSBACH. Geosci Model Dev 6:1463–1480

Haese B (2014) A new hydrogen-isotope approach to understand North African monsoon changes in the Holocene. Dissertation, Alfred-Wegener-Institut für Polar- und Meeresforschung, Universität Bremen

Kröpelin S, Verschuren D, Lezine AM, Eggermont H, Cocquyt C, Francus P, Cazet JP, Fagot M, Rumes B, Russell JM, Darius F, Conley DJ, Schuster M, von Suchodoletz H, Engstrom DR (2008) Climate-driven ecosystem succession in the Sahara: the past 6,000 years. Science 320:765–768

Laskar J, Robutel P, Joutel F, Gastineau M, Correia ACM, Levrard B (2004) A long-term numerical solution for the insolation quantities of the Earth. Astron Astrophys 428:261–285

Mulitza S, Heslop D, Pittauerova D, Fischer HW, Meyer I, Stuut JB, Zabel M, Mollenhauer G, Collins JA, Kuhnert H, Schulz M (2010) Increase in African dust flux at the onset of commercial agriculture in the Sahel region. Nature 466:226–228

Sachse D, Billault I, Bowen GJ, Chikaraishi Y, Dawson TE, Feakins SJ, Freeman KH, Magill CR, McInerney FA, van der Meer MTJ, Polissar PJ, Robins RJ, Sachs JP, Schmidt HL, Sessions AL, White JWC, West JB, Kahmen A (2012) Molecular paleohydrology: interpreting the Hydrogen-Isotopic composition of lipid biomarkers from photosynthesizing organisms. Annu Rev Earth Planetary Sci 40:221–249

Schefuß E, Kuhlmann H, Mollenhauer G, Prange M, Pätzold J (2011) Forcing of wet phases in southeast Africa over the past 17,000 years. Nature 480:509–512

Tjallingii R, Claussen M, Stuut JBW, Fohlmeister J, Jahn A, Bickert T, Lamy F, Rohl U (2008) Coherent high- and low-latitude control of the northwest African hydrological balance. Nat Geosci 1:670–675

Holocene Climate Dynamics, Biogeochemical Cycles and Ecosystem Variability in the Eastern Mediterranean Sea

Gerhard Schmiedl, Fanny Adloff, Kay-Christian Emeis,
Rosina Grimm, Michal Kucera, Ernst Maier-Reimer,
Uwe Mikolajewicz, Jürgen Möbius and Katharina Müller-Navarra

Abstract To understand the processes leading to the formation of Holocene sapropel S1 in the Eastern Mediterranean Sea, we integrated results from regional ocean-biogeochemical general circulation model experiments with biogeochemical and micropaleontological proxy records. Sapropel S1 formed during the Holocene insolation maximum, when strong Aegean north winds (Etesian) caused enhanced downwelling and mixing of warm surface waters in the Cretan and western Levantine seas accounting for the complex sea-surface temperature pattern derived from planktonic foraminiferal transfer functions. Our results support a scenario where sufficient organic matter for sapropel formation is buried under oligotrophic conditions in an anoxic water column refuting the "high-productivity" hypothesis. We reconstructed a synchronous shift in the state of deep-sea benthic ecosystems, documenting a rapid expansion of dysoxic to anoxic conditions with onset of S1

Ernst Maier-Reimer deceased.

G. Schmiedl (✉) · K.-C. Emeis · J. Möbius · K. Müller-Navarra
Center for Earth System Research and Sustainability, University of Hamburg,
Hamburg, Germany
e-mail: gerhard.schmiedl@uni-hamburg.de

F. Adloff · R. Grimm · E. Maier-Reimer · U. Mikolajewicz
Max Planck Institute for Meteorology, Hamburg, Germany

K.-C. Emeis
Helmholtz-Zentrum Geesthacht—Centre for Materials and Coastal Research,
Geesthacht, Germany

F. Adloff
Météo-France, CNRM-GAME, Toulouse, France

M. Kucera
MARUM—Center for Marine Environmental Sciences and Faculty of Geosciences,
University of Bremen, Bremen, Germany

© The Author(s) 2015   115
M. Schulz and A. Paul (eds.), *Integrated Analysis of Interglacial Climate
Dynamics (INTERDYNAMIC)*, SpringerBriefs in Earth System Sciences,
DOI 10.1007/978-3-319-00693-2_19

deposition. The recovery of benthic ecosystems during the terminal S1 phase was controlled by increasingly deeper convection and re-ventilation over a period of approximately 1,500 years.

Keywords Paleoceanography · Paleoclimatology · Numerical climate modeling · Biogeochemistry · Micropaleontology · Stable isotope geochemistry · Eastern Mediterranean Sea · Holocene · Sapropel S1

1 Introduction

The present Mediterranean Sea has a negative water balance, which drives an anti-estuarine circulation and provokes oligotrophic conditions in the surface mixed layer of the Eastern Mediterranean Sea (EMS) Krom et al. (2010). Atlantic surface water depleted in nutrients enters the Mediterranean Sea through the Strait of Gibraltar and is partly compensated by the outflow of Mediterranean intermediate water that exports nutrients mineralized from sinking organic matter. Today, high evaporation rates and cooling of surface waters trigger deep winter convection and initiate the formation of oxygen-rich deep and intermediate water masses in various sites of the basins.

Since the late Miocene, the periodic deposition of organic matter-rich sediments, so-called sapropels, documents the repeated interruption of EMS deep-water ventilation. The timing of these events is closely linked to Northern hemisphere summer insolation maxima and is attributed to phases of enhanced riverine freshwater runoff, surface-water warming and increased stratification at the base of the mixed layer ("stagnation" hypothesis) (Lourens et al. 1996; Cramp and O´Sullivan 1999). Controversy still exists on the roles of productivity and preservation in causing elevated organic carbon accumulation in sapropels ("high-productivity" hypothesis), and the required nutrient sources (Emeis and Weissert 2009).

We combined simulations with a regional ocean circulation model coupled to a biogeochemical model with a compilation of proxy data to examine the processes and spatio-temporal environmental changes of sapropel S1 formation during the early Holocene (from 10.2 thousand years (ka) before present (BP) to 6.4 ka BP) and to test the existing hypotheses.

2 Materials and Methods

We used the ocean general circulation model MPIOM in a regional configuration for the Mediterranean (Mikolajewicz 2011; Adloff et al. 2011) coupled to the marine biogeochemical model HAMOCC (Grimm 2012; Ilyina et al. 2013) forced by atmospheric fluxes from global Earth system model simulations for the 9 ka BP time slice. A Holocene insolation maximum (HIM) baseline simulation represented the pre-sapropel well-ventilated conditions (Adloff et al. 2011).To assess the sensitivity to increased freshwater input, a set of perturbation experiments was implemented to modify the baseline simulation. These include: the opening of the Bosporus, strong increase in Nile and Po runoff, freshening from the inflowing Atlantic water, and increased precipitation. We implemented an age tracer in the model to infer the stability and duration of EMS deep-water stagnation caused by each perturbation (Adloff 2011; Grimm 2012).

To validate the model results, we compiled proxy data from existing studies and generated a new multi-proxy dataset. We produced two independent sea-surface temperature (SST) records using planktonic foraminiferal transfer functions (Adloff et al. 2011) and alkenones. In addition, we estimated ventilation and organic matter flux using benthic foraminiferal assemblages and stable carbon isotopes (Schmiedl et al. 2010). Finally, using stable nitrogen isotope and organic carbon data, we evaluated productivity and carbon sequestration (Möbius et al. 2010). Monomeric distribution of amino acids allowed us to estimate the state of organic matter decomposition expressed in the degradation index (DI) (Möbius et al. 2010). For the quantification of oxygen changes in benthic ecosystems, we applied an OI that is based on the ratio between high- and low-oxygen benthic foraminiferal indicator taxa and the benthic foraminiferal diversity (Schmiedl et al. 2010).

3 Key Findings

During the HIM, our model simulated an enhanced seasonal cycle with a homogenous winter cooling of the water column and a summer warming restricted to the top few meters. The enhanced summer insolation induced a surface warming with well-defined spatial patterns of subsurface warming/cooling in the Cretan and western Levantine seas. An initial comparison of the simulated SST for the HIM and a reconstruction based on planktonic foraminiferal transfer functions showed a poor agreement especially for summer, when the vertical temperature gradient is steep. The model-proxy agreement improved considerably when the proxy data were calibrated to the entire modern habitat depth ranges of surface-dwelling planktonic foraminifers (Fig. 1). This implies that planktonic foraminifers in this case predominantly record an integrated upper ocean signal. The dynamical explanation for the regional contrasts of subsurface temperatures during the HIM summers is the wind-driven transport of warm surface waters due to strengthened northerly Aegean (Etesian) winds, amplified by enhanced vertical mixing. This leads to subsurface warming in areas of convergence, e.g., in the Cretan and western Levantine seas, and to subsurface cooling in areas of divergence. Such a process may be characteristic for time intervals of enhanced summer insolation in the EMS during the late Neogene and Quaternary (Adloff et al. 2011).

The compiled biogeochemical records and benthic foraminiferal successions across sapropel S1 from various sediment cores suggest that the accumulation of organic matter in the abyssal parts of the basins occurred under low-productive

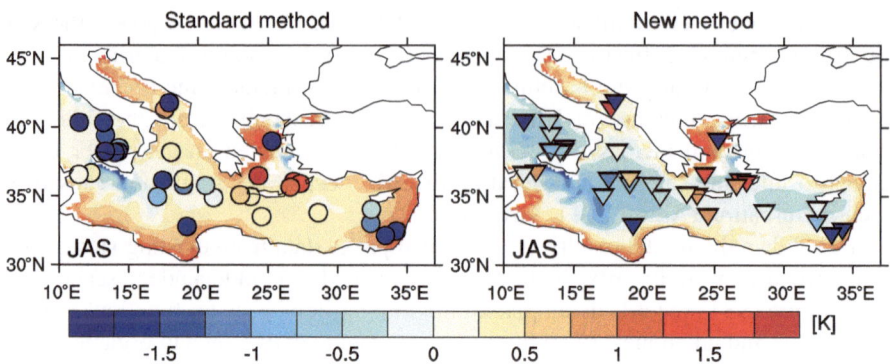

Fig. 1 Model/proxy comparison of temperature signal anomalies (9 ka BP vs. CTRL) during summer (JAS). The *left plot* displays SSTs comparison (standard method), the *right plot* displays a comparison of the T_{0-30} signal (integrated temperatures over the depth interval between 0 and 30 m). The new method consisted in recalculating reconstructions for the temperature signal including the subsurface to consider the living depth of the foraminifera used as proxy. This calculation was based on the existing SST reconstructions. Both former SSTs reconstruction (*dots on the left plot*) and new T_{0-30} reconstruction (*triangle on the right plot*) are compared to model results (*background color*) for the HIM (9 ka BP) versus CTRL (Adloff et al. 2011)

conditions, thus questioning the long-standing "high-productivity" hypothesis. Our results show that the differential degrees of organic matter diagenesis controlled the spatial and temporal variations in $\delta^{15}N$ instead of increased nitrogen availability and primary productivity (Möbius et al. 2010). This implies that the enhanced accumulation of organic carbon and observed $\delta^{15}N$ depletion in sapropels is caused by the absence of oxygen in deep waters. This scenario is in accordance with the establishment of deep-sea faunas dominated by epibenthic taxa during the earliest Holocene, a few centuries before the onset of sapropel S1, indicating the establishment of oligotrophic conditions. Our results suggest that present-day organic carbon burial fluxes, as estimated from sediment trap studies, could be sufficient to create a sapropel layer under anoxic deep-water conditions (Fig. 2a).

Records of the foraminifer-based oxygen index from the EMS imply an exponential decrease of average oxygen levels with increasing water depth across the S1 interval, suggesting a basin-wide shallowing of vertical convection. Our results reveal rapid regime shifts in bathyal and abyssal benthic ecosystems from a high-diverse state to a low-diverse or even azoic state almost synchronously with onset of S1 deposition (Fig. 2b). This suggests a rapid vertical and horizontal propagation of dysoxic to anoxic conditions, supporting previous geochemical evidence (De Lange et al. 2008). This result also suggests the ecological importance of oxygen thresholds. The recovery of bathyal deep-sea benthic ecosystems of the EMS starts around 8 ka BP at upper bathyal depths and is controlled by subsequently deeper convection and re-ventilation over a period of approximately 1,500 years (Fig. 2b) (Schmiedl et al. 2010).

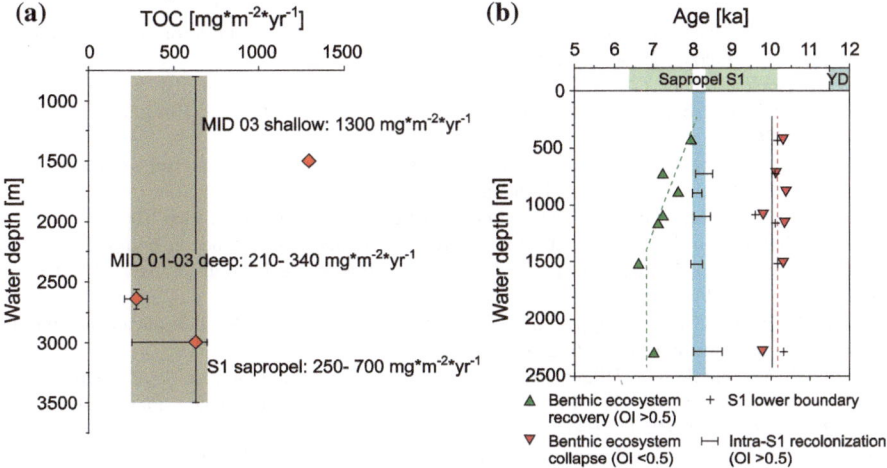

Fig. 2 **a** Total organic carbon (TOC) accumulation rates in sapropel S1 (observed range indicated by *grey bar*) (De Lange et al. 2008; Möbius et al. 2010) compared to TOC flux calculated from EMS sediment traps (MID 03 shallow and MID 01–03 deep) (Möbius et al. 2010). Data indicate that present day export production would suffice to create a S1-like sapropel, under anoxic deep water conditions. **b** Temporal succession of major benthic ecosystem changes at different water depth, associated with sapropel S1 (modified from Schmiedl et al. 2010). The ecosystem changes are indicated by passing of an OI threshold value of 0.5. The *dashed lines* display mean values of ecosystem shifts, the *solid line* marks the average for the onset of sapropel formation. The time window of EMS sapropel S1 formation and its interruption around 8.2 ka BP are indicated. *YD* Younger Dryas

By combining proxy data and modeling results we were able to explore the physical and biogeochemical processes and time-scales leading from enhanced insolation and seasonality and local hydrological perturbations to the formation of anoxia in the EMS and its regional environmental expression during the early Holocene. Based on this integrative approach we reconciled previously contradictory results, particularly concerning the observed SST patterns. We reconstructed a complex regional pattern of surface water mixing, rapid expansion of dysoxia and anoxia within a few centuries, and demonstrated that sapropel S1 was deposited under oligotrophic conditions refuting the "high-productivity" hypothesis.

References

Adloff F, Mikolajewicz U, Kucera M, Grimm R, Maier-Reimer E, Schmiedl G, Emeis KC (2011) Upper ocean climate of the Eastern Mediterranean Sea during the Holocene insolation maximum—a model study. Clim Past 7:1103–1122. doi:10.5194/cp-7-1103-2011

Adloff F (2011) Early Holocene Eastern Mediterranean ocean climate and the stability of its overturning circulation. Rep Earth Sys Sci 107:1–157

Cramp A, O'Sullivan G (1999) Neogene sapropels in the Mediterranean: A review. Mar Geol 153:11–28. doi:10.1016/S0025-3227(98)00092-9

De Lange GJ, Thomson J, Reitz A, Slomp CP, Speranza Principato M, Erba E, Corselli C (2008) Synchronous basin-wide formation and redox-controlled preservation of a Mediterranean sapropel. Nat Geosci 1:606–610. doi:10.1038/ngeo283

Emeis KC, Weissert H (2009) Tethyan-Mediterranean organic carbon-rich sediments from Mesozoic black shales to sapropels. Sedimentology 56:247–266. doi:10.1111/j.1365-3091. 2008.01026.x

Grimm R (2012) Simulating the early Holocene Eastern Mediterranean sapropel formation using an ocean biogeochemical model. Rep Earth Sys Sci 123:1–156

Ilyina T, Six K, Segschneider J, Maier-Reimer E, Li H, Núñez-Riboni I (2013) The global ocean biogeochemistry model HAMOCC: Model architecture and performance as component of the MPI-Earth System Model in different CMIP5 experimental realizations. J Adv Model Earth Sys. doi:10.1002/jame.20017

Krom MD, Emeis KC, Van Cappellen P (2010) Why is the Eastern Mediterranean phosphorus limited? Prog Oceanogr 85:236–244

Lourens LJ, Antonarakou A, Hilgen FJ, Van Hoof AAM, Vergnaud-Grazzini C, Zachariasse WJ (1996) Evaluation of the Plio-Pleistocene astronomical timescale. Paleoceanography 11:391–413. doi:10.1029/96PA01125

Mikolajewicz U (2011) Modeling Mediterranean ocean climate of the last glacial maximum. Clim Past 7:161–180. doi:10.5194/cp-7-161-2011

Möbius J, Lahajnar N, Emeis KC (2010) Diagenetic control of nitrogen isotope ratios in Holocene sapropels and recent sediments from the Eastern Mediterranean Sea. Biogeosciences 7:3901–3914. doi:10.5194/bg-7-3901-2010

Schmiedl G, Kuhnt T, Ehrmann W, Emeis KC, Hamann Y, Kotthoff U, Dulski P, Pross J (2010) Climatic forcing of eastern Mediterranean deep-water formation and benthic ecosystems during the past 22,000 years. Quatern Sci Rev 29:3006–3020. doi:10.1016/j.quascirev.2010.07.002

Environmental and Climate Dynamics During the Last Two Glacial Terminations and Interglacials in the Black Sea/Northern Anatolian Region

Helge W. Arz, Lyudmila S. Shumilovskikh, Antje Wegwerth, Dominik Fleitmann and Hermann Behling

Abstract This study provides the first detailed multi-proxy paleoenvironmental reconstructions of changes in the aquatic and terrestrial ecosystems during the Holocene, Eemian and the last two glacial/interglacial transitions (Terminations I and II) by studying sediment cores from the southeastern Black Sea and stalagmite studies from Sofular Cave in northwestern Anatolia. The terrestrial proxies document gradual changes from late glacial cold/arid conditions in northern Anatolia, dominated by steppe vegetation, to warm/humid forest dominated landscapes characteristic for interglacial periods. The Holocene and Eemian, however, developed differently, with warmer and moister conditions prevailing during the Eemian. Major fluctuations in the hydrological state of the Black Sea are closely linked to changes of terrestrial environments. Disrupted by large melt water pulses from the disintegrating Fennoscandian Ice Sheet, the limnic glacial Black Sea environment becomes more productive during the postglacial warming. Global sea-level rise finally reconnects the hydrological increasingly active Black Sea basin with the Mediterranean Sea leading to the development of marine, for the Eemian even fully marine, conditions with a stratified water column and sapropelic sedimentation.

Keywords Black sea · Eemian · Holocene · Glacial-interglacial transitions · Paleoenvironment · Paleoclimatology · Palynology · Geochemistry

H.W. Arz (✉) · A. Wegwerth
Leibniz Institute for Baltic Sea Research Warnemünde, Rostock-Warnemünde, Germany
e-mail: helge.arz@io-warnemuende.de

L.S. Shumilovskikh · H. Behling
Department of Palynology and Climate Dynamics, Georg-August-University Göttingen, Göttingen, Germany

D. Fleitmann
Institute of Geological Sciences and Oeschger Centre for Climate Change Research, University of Bern, Bern, Switzerland

D. Fleitmann
Department of Archaeology, School of Human and Environmental Sciences, University of Reading, Reading, UK

© The Author(s) 2015
M. Schulz and A. Paul (eds.), *Integrated Analysis of Interglacial Climate Dynamics (INTERDYNAMIC)*, SpringerBriefs in Earth System Sciences, DOI 10.1007/978-3-319-00693-2_20

1 Introduction

Due to its narrow/shallow Bosporus Strait connection to the Mediterranean, the Black Sea is a well stratified marginal sea since about 8,000 years. However, on longer timescales the Black Sea is characterized by transient environments critically depending on sea level and climate changes on glacial/interglacial time scales with marine conditions during warm interglacials and freshwater/brackish conditions during glacials (Badertscher et al. 2011; Shumilovskikh et al. 2012). Particularly during the last glacial terminations, shifts from limnic to almost fully marine conditions were accompanied by major changes on the bordering terrestrial environments that are presently under the influence of Atlantic/Mediterranean and central to eastern European climates. While the Last Glacial Maximum to Holocene transition received some attention in the western Black Sea (e.g., Bahr et al. 2008) the lack of adequate sediments and poorly dated terrestrial archives inhibited a detailed land-ocean comparison for older glacial/interglacial transitions (also known as terminations) and interglacials.

New sediment cores from the SE Black Sea, covering the last 134 thousand years (ka), provide detailed paleoceanographic/paleolimnologic data of environmental changes in the marine/limnic Black Sea as well as records of vegetation dynamics and changing precipitation regimes in the Anatolian hinterland. Stalagmites from Sofular Cave, northwestern Anatolia, serve as long complementary terrestrial paleorecords and help to construct a precise chronology for the sediment cores. Combining both archives allows us to quantify land-ocean interaction on the regional scale more precisely and helps us to consider the impact of hemispheric/North Atlantic climate and global sea level changes during the Holocene, Eemian and the last two teminations (I + II).

2 Materials and Methods

The speleothems from Sofular cave (41° 25′ N, 31° 56′ E) in northwestern Turkey were precisely Uranium-series dated and their $\delta^{18}O$ and $\delta^{13}C$ records primarily reflect changes in the $\delta^{18}O$ of the Black Sea-surface water and soil productivity/vegetation density above the cave, respectively (Fleitmann et al. 2009; Badertscher et al. 2011; Göktürk et al. 2011). The sediment cores 22-GC3 (42° 13.53′ N, 36° 29.55′ E, 838 m bsl), 22-GC8 (42° 13.53′ N, 36° 29.59′ E, 847 m bsl), and 24-GC3 (41° 28.66′ N, 37° 11.68′ E, 208 m bsl) were collected in 2007 during the RV Meteor cruise M72/5 in the south-eastern Black Sea (Fig. 1). Sedimentological, geochemical, and palynological data were obtained on high-resolution sample series from the core intervals of interest, namely the Holocene, Eemian and the last two glacial/interglacial transitions. Elemental analyzes, x-ray fluorescence scanning (XRF), and sedimentological analyses (e.g., counting of large >150 µm detrital particles transported by ice, IRD) were used to characterize changes in the

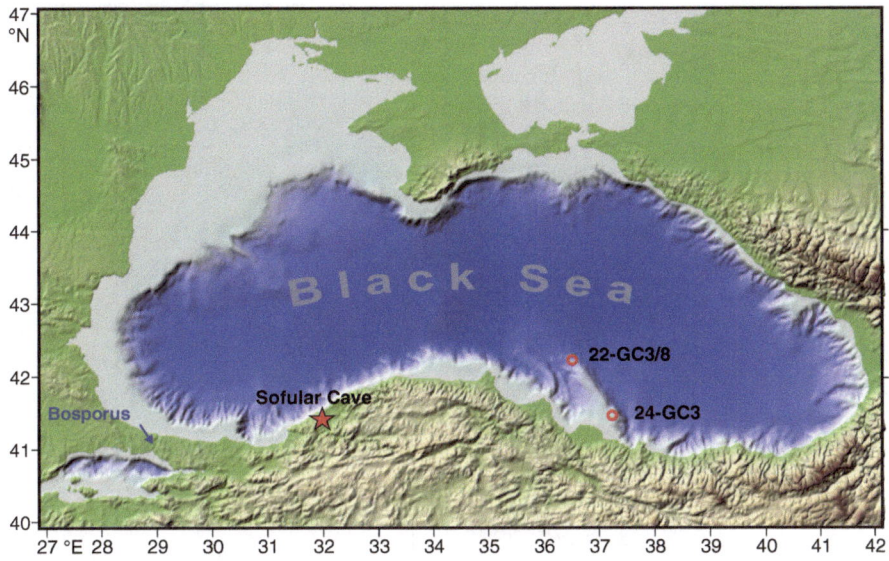

Fig. 1 Location of Sofular cave in northwestern Turkey and of the sediment cores 22-GC3/8 and 24-GC3 in the southeastern Black Sea

depositional environment (Nowaczyk et al. 2012). The stable oxygen isotope composition determined on ostracod shells indicates major hydrographic changes in the basin and was used for stratigraphic purposes. Palynological analyzes were used to evaluate vegetation changes in the hinterland (Shumilovskikh et al. 2012) while dinoflagellate cyst concentrations were used as proxy for changes in aquatic primary productivity and cyst composition for reconstructing surface salinity and temperature (Shumilovskikh et al. 2013a). The stratigraphy of the cores includes radiocarbon dates, tephra dates, paleomagnetic records, correlations of the $\delta^{18}O$ ostracod/bivalve record with the uranium-series dated $\delta^{18}O$ speleothem record from Sofular Cave, proxy matching to the Greenland $\delta^{18}O$ NGRIP ice core record, and palynological correlations to pollen records from the Mediterranean region (Nowaczyk et al. 2012; Shumilovskikh et al. 2012, 2013a).

3 Key Findings

The last two glacial terminations and subsequent interglacial periods (Holocene and Eemian) show largely similar patterns of environmental transitions in Northern Anatolia and the southern Black Sea (Fig. 2). Disconnected from the Mediterranean Sea, the Black Sea was characterized by freshwater/brackish dinocyst assemblages during the ending glacials and abundant IRD suggesting coastal ice formation during extreme winters (Fig. 2e,f; Nowaczyk et al. 2012; Shumilovskikh et al. 2012, 2013a).

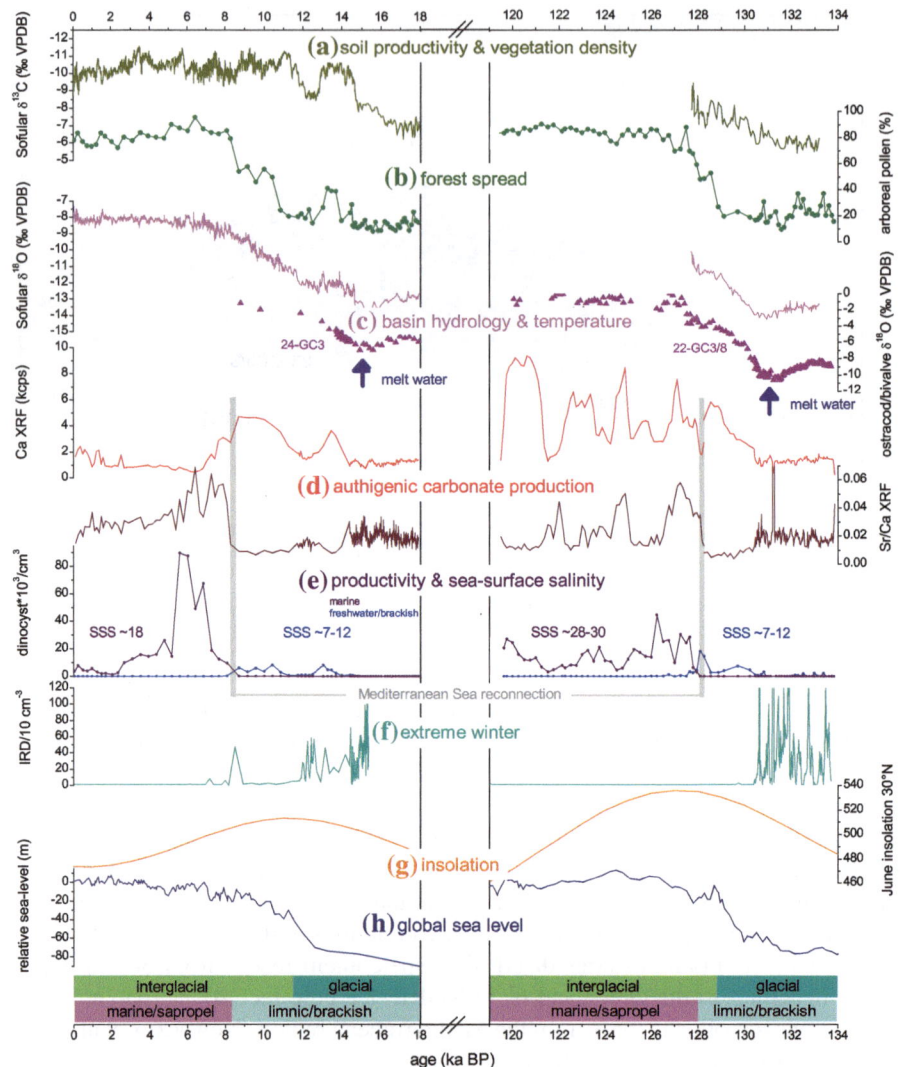

Fig. 2 Comparison of Terminations I and II based on terrestrial and aquatic proxies from the Sofular cave in northwestern Anatolia and sediment cores from the southeastern Black Sea: **a** soil productivity and vegetation density as estimated from stalagmite $\delta^{13}C$ (Fleitmann et al. 2009); **b** arboreal pollen indicating spread of forests in Northern Anatolia (Shumilovskikh et al. 2012, 2013b); **c** $\delta^{18}O$ isotopes from Sofular Cave (Badertscher et al. 2011) and from ostracods/bivalves in the Black Sea cores 22-GC3/8 (Shumilovskikh et al. 2013a) and 24-GC3 reflecting changes in the basins hydrology (*arrows*) indicate postglacial melt water events (Badertscher et al. 2011); **d** Ca XRF reflect changes in the carbonate content of the sediment (Shumilovskikh et al. 2012) with the Sr/Ca XRF ratio indicating its calcitic or aragonitic composition; **e** dinocyst concentrations indicating Black Sea productivity changes (Shumilovskikh et al. 2012), and dinocyst composition reflecting sea-surface salinity (SSS) (Shumilovskikh et al. 2013a); **f** ice-rafted debris (IRD) suggesting coastal ice formation during extreme winters (Nowaczyk et al. 2012); **g** June insolation at 30°N; **h** global sea level changes (Rohling et al. 2009)

North Anatolian speleothem $\delta^{18}O$ records (Fleitmann et al. 2009; Badertscher et al. 2011) and ostracod $\delta^{18}O$ values from the Black Sea itself unison document the early Fennoscandian Ice Sheet disintegration initially routing its melt water to the Caspian and Black Seas (Fig. 2c; Wegwerth et al. 2014). The postglacial warming of the Black Sea lake starting at ~ 130.5 and ~ 14.7 ka BP for Termination II (T II) and Termination I (T I), respectively, is accompanied by authigenic calcite precipitation (Fig. 2d) induced by augmented phytoplankton blooms during the warm season indicated e.g., by dinocyst concentrations (Fig. 2e). While T II shows steadily increasing carbonate precipitation, a prominent interruption during the Younger Dryas characterizes T I. With the connection to the Mediterranean Sea related to the postglacial sea-level rise (Eemian at ~ 128 ka BP and Holocene at ~ 8.3 ka BP), calcite precipitation ceases and organic-rich sedimentation under anoxic conditions—the sapropel formation—starts. Eemian dinocyst assemblages differ considerably from the Holocene showing the presence of fully marine species. An increased inflow of Mediterranean water triggered by a ~ 6 m higher-than-today global sea level and higher temperatures at the beginning of the Eemian are likely explaining higher-than-present sea-surface salinity of about 28–30 psu (Fig. 2e; Shumilovskikh et al. 2013a).

Terrestrial proxies, presented by pollen and speleothem $\delta^{13}C$ values, indicate open steppe and steppe-forest landscapes typical for cold/arid conditions at the end of both glacial periods and spread of forests and increased soil productivity under warm/humid conditions during both interglacials (Fig. 2a,b). Initial warming started at 130.4 ka BP during TII and at 14.7 ka BP during TI, and is clearly reflected by increase of arboreal and especially oak and pine pollen (Shumilovskikh et al. 2012, 2013b) and more negative speleothem $\delta^{13}C$ values (Fleitmann et al. 2009). The onset of interglacials is indicated by arboreal pollen exceeding 50 % (Eemian at ~ 128.8 ka BP and Holocene at ~ 10.6 ka BP) and is characterized by semi-arid conditions as revealed by the dominance of pines and oaks. Spread of typical euxinian beech forests during both interglacial periods coincides with the Mediterranean inflow into the Black Sea basin, suggesting establishment of temperate climatic conditions after 128 and 8.3 ka BP (Fig. 2b). High abundance of pine indicates generally cooler and dryer conditions during the Holocene (Shumilovskikh et al. 2012), whereas warm-temperate forests dominate the Eemian and were possibly caused by a more pronounced northern summer insolation maximum (Fig. 2g; Shumilovskikh et al. 2013b). The early Holocene humidity increase as indicated by the arboreal pollen record (Fig. 2b) is slightly later when compared to records from the eastern Mediterranean region. These delays are known from different parts of Anatolia and were potentially attributed to changes in precipitation seasonality. Less densely vegetated landscapes at the end of the Holocene suggest dryer climate and/or increased anthropogenic activity (Shumilovskikh et al. 2012).

The terrestrial and aquatic environments at the southern rim of the Black Sea demonstrably experienced large changes in the course of the last two terminations and the succeeding interglacials. Shifting precipitation/evaporation regimes largely relate to a postglacial northward retreat of the polar frontal systems and regional warming. However, the hydrological regime of the Black Sea was punctuated by distinct events, i.e., the late glacial massive melt water discharges and the early interglacial reconnections to the Mediterranean Sea pacing the particular timing and amplitude of the observed environmental changes (Wegwerth et al. 2014). Further work is needed to reconcile the impact of the Black Sea system on the eastern Mediterranean Sea. Moreover, detailed Black Sea studies on the transitions from marine to limnic environments during the late interglacial states are required.

References

Badertscher S, Fleitmann D, Cheng H, Edwards RL, Göktürk OM, Zumbühl A, Leuenberger M, Tüysüz O (2011) Pleistocene water intrusions from the Mediterranean and Caspian Seas into the Black Sea. Nat Geosci 4:236–239. doi:10.1038/ngeo1106

Bahr A, Lamy F, Arz HW, Major C, Kwiecien O, Wefer G (2008) Abrupt changes of temperature and water chemistry in the late Pleistocene and early Holocene Black Sea. Geochem Geophys Geosyst. doi:10.1029/2007GC001683

Fleitmann D, Cheng H, Badertscher S, Edwards RL, Mudelsee M, Göktürk OM, Fankhauser A, Pickering R, Raible CC, Matter A, Kramers J, Tüysüz O (2009) Timing and climatic impact of Greenland interstadials recorded in stalagmites from northern Turkey. Geophys Res Lett. doi:10.1029/2009GL040050

Göktürk OM, Fleitmann D, Badertscher S, Cheng H, Edwards RL, Leuenberger M, Fankhauser A, Tüysüz O, Kramers J (2011) Climate on the southern Black Sea coast during the Holocene: implications from the Sofular Cave record. Quaternary Sci Rev 30:2433–2445

Nowaczyk NR, Arz HW, Frank U, Kind J, Plessen B (2012) Dynamics of the Laschamp geomagnetic excursion from the Black Sea sediments. Earth Planet Sci Lett 351–352:54–69. doi:10.1016/j.epsl.2012.06.050

Rohling EJ, Grant K, Bolshaw M, Roberts AP, Siddall M, Hemleben Ch, Kucera M (2009) Antarctic temperature and global sea level closely coupled over the past five glacial cycles. Nat Geosci 2:500–504. doi:10.1038/NGEO557

Shumilovskikh LS, Tarasov P, Arz HW, Fleitmann D, Marret F, Nowaczyk N, Plessen B, Schlütz F, Behling H (2012) Vegetation and environmental dynamics in the southern Black Sea region since 18 kyr BP derived from the marine core 22-GC3. Palaeogeogr Palaeocl 337–338:177–193. doi:10.1016/j.palaeo.2012.04.015

Shumilovskikh LS, Marret F, Fleitmann D, Arz HW, Nowaczyk N, Behling H (2013a) Eemian and Holocene sea-surface conditions in the southern Black Sea: organic-walled dinoflagellate cyst record from core 22-GC3. Mar Micropaleontol 101:146–160. doi:10.1016/j.marmicro.2013.02.001

Shumilovskikh LS, Arz HW, Wegwerth A, Fleitmann D, Marret F, Nowaczyk N, Tarasov P, Behling H (2013b) Vegetation and environmental changes in Northern Anatolia during 134-119 ka recorded in Black Sea sediments. Quaternary Res 80:349–360. doi:10.1016/j.yqres.2013.07.005

Wegwerth A, Dellwig O, Kaiser J, Ménot G, Bard E, Shumilovskikh L, Schnetger B, Kleinhanns IC, Wille M, Arz HW (2014) Meltwater events and the Mediterranean reconnection at the Saalian-Eemian transition in the Black Sea. Earth Planet Sci Lett 404: 124–135. doi:10.1016/j.epsl.2014.07.030

Seasonal Reconstruction of Summer Precipitation Variability and Dating of Flood Events for the Millennium Between 3250 and 2250 Years BC for the Main Region, Southern Germany

Johannes Schoenbein, Alexander Land, Michael Friedrich, Rüdiger Glaser and Manfred Kueppers

Abstract We present a millennial-long reconstruction of summer precipitation variability for the Main region (MR) in southern Germany (3250 BC to 2250 BC) based on subfossil-oak, tree-ring data. Wood-anatomical flood markers have been identified and used for dating flood events providing additional information of hydro-climatic dynamics. Reconstructed precipitation variability and flood occurrence show a noticeable shift which coincides with a decrease in replication (number of trees) of the Holocene Oak Chronology (HOC) Hohenheim around 2750 BC. The reconstruction is based on a linear climate-growth model derived from total ring width (TRW) data from living oak trees from the MR using the past approximately 130 years as a basis. The best response was identified for the precipitation sum between April 1st and July 10th providing an $r^2 = 0.31$.

Keywords Tree rings · Southern Germany · Holocene Oak Chronology (HOC) Hohenheim · Precipitation variability · Flood frequency

1 Introduction

Studies on hydro-climatic variability had been secondary to those studying past temperature variability (Wilson et al. 2013). However ambitious efforts were undertaken to reconstruct past hydro-climatic conditions (e.g. Buentgen et al. 2011; Drobyshev et al. 2011; Wilson et al. 2013). Derived from the original idea to combine the Holocene Oak Chronology (HOC) with documentary data to provide

J. Schoenbein (✉) · R. Glaser
Department of Physical Geography, Albert-Ludwigs-University Freiburg, Freiburg, Germany
e-mail: johannes.schoenbein@geographie.uni-freiburg.de

A. Land · M. Friedrich · M. Kueppers
Institute of Botany, University of Hohenheim, Stuttgart, Germany

© The Author(s) 2015
M. Schulz and A. Paul (eds.), *Integrated Analysis of Interglacial Climate Dynamics (INTERDYNAMIC)*, SpringerBriefs in Earth System Sciences,
DOI 10.1007/978-3-319-00693-2_21

for an extended calibration and verification period, we present the results of a modeled precipitation reconstruction and flood dating based on subfossil-oak, tree-ring data from 3250 BC to 2250 BC. The researched time period was chosen due to the fact that the HOC displays a noticeable drop in replication around 2750 BC and a pronounced rise in raw total ring width (TRW) since 2731 BC. As a result of the origin of the wood samples, the reconstruction emphasizes the situation of the Main region (MR) in Germany (49° 45′ N, 9° 30′ E). In addition to TRW data, which had been used for the seasonal precipitation reconstruction, wood anatomical anomalies (WAA) had been used as a proxy for strong and long-lasting flood events.

2 Materials and Methods

Chronologies from TRW from living (1876 to 2004 AD) (Land 2013) and subfossil (3250 to 2250 BC) (e.g. Friedrich et al. 2004) oak trees from the MR were used in this study. To preserve high- to mid-frequency variability, tree-ring series were individually detrended using 67 % cubic smoothing splines with 50 % frequency response cutoff (Cook and Peters 1981). Annual indices were calculated as ratios from the fitted growth curves. Variance adjusted chronologies (living and subfossil) were generated using a bi-weight robust mean (Cook 1985). Signal strength of the two chronologies was assessed using a moving window of 50 years (with 25 years overlap) of the inter-series correlation (RBar) and expressed population signal (EPS) (Wigley et al. 1984).

Additionally all living and about 20 % of the subfossil oak samples had been analyzed for characteristic changes in the wood anatomy which can be regarded as a reliable proxy for long-lasting flood events (e.g. Astrade and Begin 1997; Land 2013; Tardif and Bergeron 1997). WAA in oak trees are formed in the submerged parts of a stem. The characteristics of WAA allow for differentiation between summer and winter floods (Astrade and Begin 1997; Land 2013; Tardif and Bergeron 1997). In order to identify the exact timing of the growth response of TRW indices to the climatological trigger an algorithm programmed in MATLAB (Schoenbein 2011) was used. Based on climate data in daily resolution (e.g. temperature, precipitation, hours of sunshine, drought index) the script aggregated the data for each year altering (a) the length of the data sample used for correlation between 21 and 181 days and (b) the starting date of the sample derived from (a) between January 1st of the previous year and December 15th of the current year (Schoenbein 2011). A Pearson correlation between each generated data set and the TRW index for the MR was calculated. Highest correlations were selected and response functions were generated. A significant relation was found between TRW index and the sum of the daily precipitation from April 1st to July 10th. The linear climate-growth model derived from the correlation [$Precipitation$ (mm) $= 56.97 + 127.81 \times TRW$] had been applied to the TRW index from subfossil oak trees.

3 Key Findings

The calibration of the regression model within the full 1876 AD to 2004 AD and two split periods, the number of living oak trees, RBar and the EPS is shown in Fig. 1a, b. The corresponding patterns of TRW and measured seasonal precipitation sum for the calibration period is displayed. The reconstruction of Holocene summer precipitation variability between 3250 BC and 2250 BC is displayed in Fig. 2.

The raw TRW (Fig. 2a) of the MR dropped between 2750 BC and 2731 BC and shows a strong increase later on. This period is framed by a strong decrease in replication stretching from around 2800 BC to 2650 BC (Fig. 2b). The precipitation reconstruction (Fig. 2c) shows a drop for the period from 2746 BC to 2727 BC which did not occur at another time during the researched period. Prior to this period an elevated number of severe wet ($>2\sigma$) summers occurred which was followed by roughly 170 years where severely dry ($<-2\sigma$) summers had been reconstructed. These findings might indicate a changing hydrodynamic activity around 2750 BC. In addition a period of high flood frequency from 2991 BC to 2693 BC can be identified. 10 of 30 flood events coincided with years where the reconstruction shows high precipitation ($>1\sigma$); see Fig. 2a.

The increase in raw TRW (Fig. 2a), which is uncommon in the HOC until that time (Friedrich et al. 2004), can be interpreted as a signal for changing forest dynamics (Spurk et al. 2002). Spurk et al. (2002) explain the raw TRW increase as phase of succession which might follow a period of wetter than normal climate which suppressed germination due to a change in climate to more suitable growing

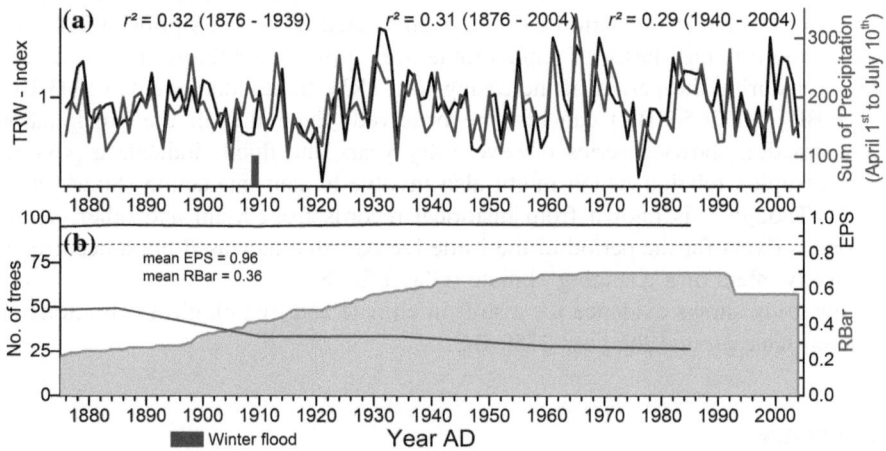

Fig. 1 Relation between TRW index and seasonal precipitation for the MR. **a** Synchronicity between TRW index (*black*) and precipitation sum of April 1st to July 10th (*blue*) for the MR over the full period (1876 AD to 2004 AD) and two split periods. Year to year precipitation variability finds good expression in TRW index. From historical data there is only one medium winter flood event known (*blue bar*, 1909 AD), however no WAA had been found for this event (see text below). **b** Number of living oak trees used to develop the TRW index (*grey filled area*), RBar (*grey*) and EPS (*black*). Values for RBar and EPS indicate high common signal strength

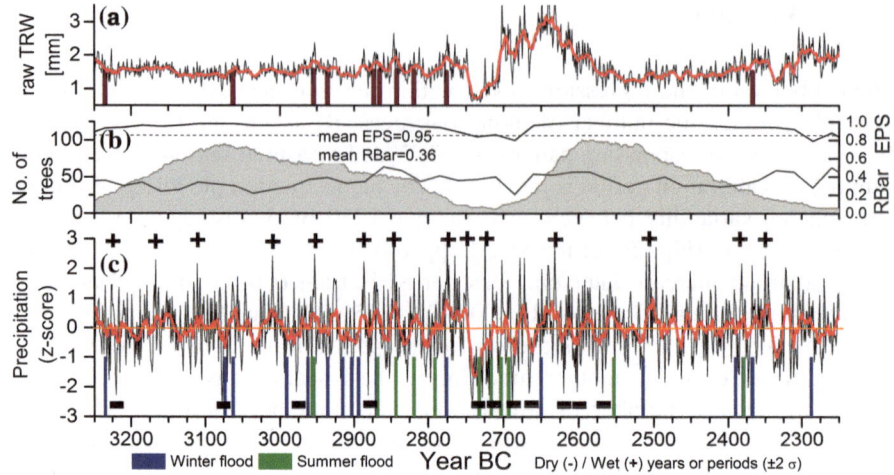

Fig. 2 Reconstructed seasonal precipitation for 3250 BC to 2250 BC. **a** Raw TRW (*black*) and 11-year moving average (*red*) of the HOC for the millennium between 3250 BC and 2250 BC of the MR. *Brown bars* indicate a coincidence of flood occurrence derived from WAA and high precipitation events (>1 sample standard deviation [σ]). **b** Number of subfossil oak trees used for TRW index development for the MR (*grey filled area*). The values for RBar (*grey*) and EPS (*black*) display high common signal strength except for a short period at 2700 BC. **c** Seasonal reconstruction of summer precipitation variability. Severe wet (>2σ) years are marked with [+] and severe dry (<−2σ) years are marked with [−]. Flood events based on WAA are marked with *colored bars*

conditions. The precipitation reconstruction for the period from 2785 BC to 2747 BC as well as an accumulation of flood events dated from WAA prior to 2747 BC would confirm the thesis of unfavorable wet growth conditions for seedlings. However during the period of increasing raw TRW three flood events (2703 BC, 2693 BC and 2650 BC) can also be observed. Even though the precipitation reconstruction shows a series of severe dry years, the floods indicate a possible more complex relation which might also involve human interaction (Spurk et al. 2002). Though it is known from historical records from Main and other central European rivers for the period of the Little Ice Age that increased flood occurrence is a likely effect of a changing climate (Glaser 2008).

Our study shows evidence for a shift in climate and hydrological conditions on regional scale around the year 2750 BC.

References

Astrade L, Begin Y (1997) Tree-ring response of *Populus tremula* L. and *Quercus robur* L. to recent spring floods of the Saone River, France. Ecoscience 4(2):232–239

Buentgen U, Tegel W, Nicolussi K, McCormick M, Frank D, Trouet V, Kaplan JO, Herzig F, Heussner KU, Wanner H, Luterbacher J, Esper J (2011) 2500 Years of European climate variability and human susceptibility. Science 331(6017):578–582. doi:10.1126/science.1197175

Cook ER (1985) A time series analysis approach to tree ring standardization. Dissertation, The University of Arizona

Cook ER, Peters K (1981) The smoothing spline: a new approach to standardizing forest interior tree-ring width series for dendroclimatic studies. Tree-Ring Bull 41:45–53

Drobyshev I, Niklasson M, Linderholm HW, Seftingen K, Hickler T, Eggertsson O (2011) Reconstruction of a regional drought index in southern Sweden since AD 1750. Holocene 21 (4):667–679. doi:10.1177/0959683610391312

Friedrich M, Remmele S, Kromer B, Spurk M, Hofmann J, Hurni JP, Kaiser KF, Küppers M (2004) The 12,480-year Hohenheim oak and pine tree-ring chronology from Central Europe—a unique annual record for radiocarbon calibration and palaeoenvironment reconstructions. Radiocarbon 46(3):1111–1122

Glaser R (2008) Klimageschichte Mitteleuropas; 1200 Jahre Wetter, Klima. Katastrophen, Primusverlag, Darmstadt

Land A (2013) Holzanatomische Veränderungen als Reaktion auf extreme Umweltereignisse in rezenten und subfossilen Eichen Süddeutschlands und deren Verifizierung im Experiment. Dissertation, University of Hohenheim

Schoenbein J (2011) Zur Rekonstruktion von Hochwasserereignissen in Europa aus holzanatomischen Parametern und historischen Quellen. Dissertation, University of Freiburg

Spurk M, Leuschner HH, Baillie MGL, Briffa KR, Friedrich M (2002) Depositional frequency of German subfossil oaks: climatically and non-climatically induced fluctuations in the Holocene. Holocene 12(6):707–715. doi:10.1191/0959683602hl583rp

Tardif J, Bergeron Y (1997) Ice-flood history reconstructed with tree-rings from the southern boreal forest limit, western Quebec. Holocene 7(3):291–300. doi:10.1177/095968369700700305

Wigley TML, Briffa KR, Jones PD (1984) On the average of value of correlated time series, with applications in dendroclimatology and hydrometeorology. J Clim Appl Meteorol 23:201–213

Wilson R, Miles D, Loader N, Melvin T, Cunningham L, Cooper R, Briffa K (2013) A millennial long March–July precipitation reconstruction for southern-central England. Clim Dyn 40 (3–4):997–1017. doi:10.1007/s00382-012-1318-z

Precipitation in the Past Millennium in Europe—Extension to Roman Times

Juan Jose Gómez-Navarro, Johannes P. Werner, Sebastian Wagner, Eduardo Zorita and Jürg Luterbacher

Abstract This project aimed at describing the evolution of precipitation and its variability over Europe and the Mediterranean over the last two millennia. We present results from dynamical downscaling, showing the added value of regional climate models in the paleoclimate context. The regional models improve the representation of precipitation patterns and variability compared to the raw global climate model output and indicate periods with warmer/drier and colder/wetter summer conditions throughout the last two millennia, including for instance the Medieval Climate Anomaly and the Little Ice Age. Additionally, based on the regional simulations pseudoproxies are generated to test the analog method and Bayesian inference. The application of the Bayesian and analog methods to pseudoproxies show reasonable skill and can be used as a statistical tools for the reconstruction of hydrological-sensitive proxy data and might be appropriate methods to be applied to real proxies in the future.

Keywords Natural climate variability · Late holocene · Regional climate modelling · Bayesian hierarchical modelling · Precipitation reconstructions · Changes in hydrological cycle

S. Wagner · E. Zorita
Institute for Coastal Research, Helmholtz-Zentrum Geesthacht, Geesthacht, Germany

J.P. Werner · J. Luterbacher
Department of Geography, University of Giessen, Giessen, Germany

J.J. Gómez-Navarro (✉)
Climate and Environmental Physics, Physics Institute and Oeschger Centre for Climate Change Research, University of Bern, Bern, Switzerland
e-mail: gomez@climate.unibe.ch

© The Author(s) 2015

133

M. Schulz and A. Paul (eds.), *Integrated Analysis of Interglacial Climate Dynamics (INTERDYNAMIC)*, SpringerBriefs in Earth System Sciences, DOI 10.1007/978-3-319-00693-2_22

1 Introduction

In the climate system, the hydrological cycle is a key component of many processes important for the radiative balance and heat transport related to cloudiness (e.g., Tang et al. 2012; Gagen et al. 2011) and the exchange of sensible and latent heat with the Earth's surface (e.g., Seneviratne et al. 2010).

Water supply for instance in the Mediterranean region and agriculture in semi-arid regions is the major climatic factor that impact human societies and ecosystems. For longer time scales paleoclimatology offers the possibility to study changes in the past hydrological cycle before meteorological measurements are available.

A previous study (Pauling et al. 2006) reconstructed spatial precipitation fields over Europe applying linear methods (Principal Component Regression or simple ordinary least square regression). This project aimed at reconstructing the variations of spatially resolved European precipitation fields over the past two millennia by implementing novel statistical reconstruction approaches [Bayesian Hierarchical modeling (BHM) and analog method] using multiproxy data and regional climate modeling (dynamical downscaling). The ultimate goals were to analyze the spatio-temporal variability of past precipitation changes over continental Europe (including parts of the Mediterranean) and to assess the associated uncertainties on decadal-to-centennial time scales. A 2000-year spatio-temporal highly resolved reconstruction represents the backdrop to frame present and projected anthropogenic climate changes.

The application of regional modeling combined with the BHM (Tingley and Huybers 2010; Werner et al. 2013) and analog method (Franke et al. 2011) allow better estimations of past precipitation variations. The output of the regional model offers the possibility to test statistical methods such as the BHM and analog method for the generation of pseudoproxies with the aim to reconstruct spatially resolved precipitation patterns. In the following, examples for two European regions, the larger Alpine area and central Europe representing different levels of spatial heterogeneity, will be used to characterize the skill of the statistical and precipitation and its evolution over the last 2,000 years.

2 Materials and Methods

The resolution of global climate models (GCMs) used within the paleoclimatic context is still very coarse with grid spacings in the order of a few hundreds of kilometers. To analyze precipitation variability, the European climate over the past two millennia is therefore downscaled with high resolution regional climate models (RCMs). The finer horizontal resolution in the order of 50 km with a model time step of typically three minutes allows a better representation of precipitation processes, for instance related to the formulation of convective precipitation.

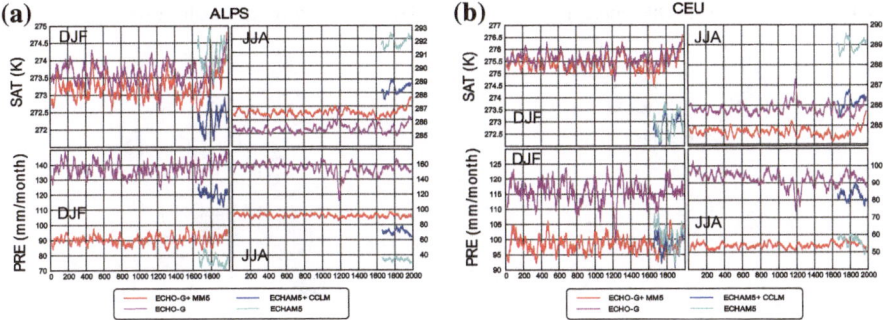

Fig. 1 Series of SAT (in K) and precipitation (in mm/month) spatially averaged for the Alps (**a**) and Central Europe (**b**). Winter and summer is presented separately, the four series represent the two GCM runs (*cyan* and *pink*) and the two regional models used to downscale these runs (*blue* and *red*). All series have been smoothed with a 31-year running mean

The higher resolution of the RCM is of particular importance because hydrological sensitive proxy data that will be used for comparison and provide the basis for empirical reconstructions are usually only representative of their specific regional-to-local setting.

Two different regional models are used here: MM5 (Gomez-Navarro et al. 2013) and CCLM (Rockel and Geyer 2008) are driven at the domain boundaries by simulations with global atmosphere-ocean general circulation models (AOGCM). The CCLM simulation is forced at its lateral boundaries with the Earth System Model (ESM) ECHAM5-MPIOM-JSBACH (Jungclaus et al. 2010) for the period 1645–2000. The combined ECHAM5+CCLM model is driven by prescribed changes in orbital, solar and volcanic forcings as well as changes in land use. Changes in atmospheric greenhouse gases are interactively calculated by the sub-model of the carbon-cycle embedded in the ESM itself.

A second regional simulation is performed with the regional climate model MM5, forced with the AOGCM ECHO-G (González-Rouco et al. 2003; Zorita et al. 2005) for the period 1–1998. Due to the simpler structure of ECHO-G, only changes in orbital, solar and greenhouse gas forcings are considered. In the combined ECHO-G+MM5 simulation, greenhouse trace gases are prescribed based on estimations from ice core data (Flückiger et al. 2002). For comparison also the raw output of the driving GCMs ECHAM5 and ECHO-G will be presented in addition to the numerically downscaled (ECHAM5+CCLM and ECHO-G+MM5) for the sample regions in the results section (cf. Fig. 1).

We use the downscaled ECHO-G+MM5 output in a Pseudo-Proxy Experiment [PPE, cf. Rutherford et al. (2005) and Smerdon (2012) for a review] to assess two reconstruction methods for precipitation over Europe and the Mediterranean region: the analog method (Franke et al. 2011) and BARCAST (Tingley and Huybers 2010; Werner et al. 2013) and their ability to reconstruct Winter (DJF) precipitation sums. PPEs have the advantage that the experimental conditions, such as proxy distribution and noise, can be tuned and the true target field is known. This enables

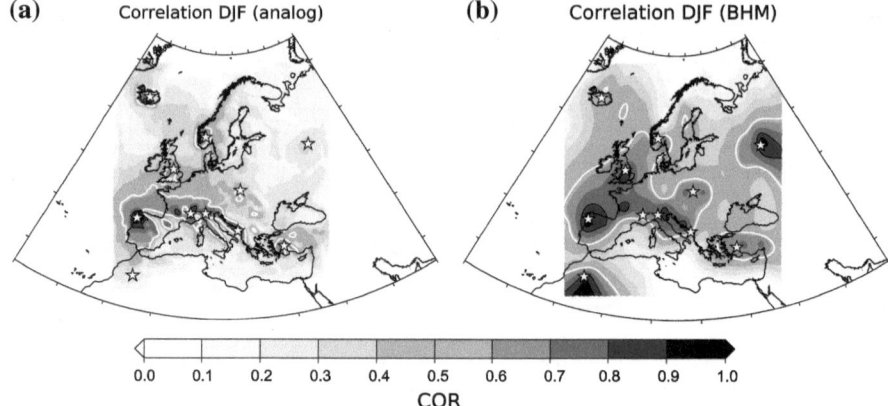

Fig. 2 Correlation between the reconstructions and the actual ECHO-G+MM5 model run for winter season precipitation. **a** Analog method, **b** BARCAST. *White stars* indicate the location of the pseudoproxies used to perform the reconstruction. The period used to calculate the correlations is 1–1998. For BARCAST the reconstruction period was 1–1850, with a calibration interval of 1851–1998. Values inside the area delimited by *white contours* are statistically significant with $p < 0.05$. The significance level is estimated by Monte-Carlo simulations bootstrapping the precipitation series

systematic comparison of different reconstruction methods. In a PPE, precipitation sums simulated by a GCM or RCM are used as a basis to construct artificial proxy data (pseudoproxies) at a few locations by distorting the time series with noise. The kind of noise is based on the signal-to-noise ratio (SNR) between the proxy time series and the observational variable under consideration. The typical SNR for real world proxies is in the range between 0.25 and 0.5 (by standard deviation) (Smerdon 2012). For our specific PPE, eleven locations are arbitrary selected as a set of pseudoproxies. To better mimic real-world proxy data, white noise could be added on the pseudoproxies at those locations, although in this case we started by using perfect information (noise free case). Then, the two reconstruction methods are applied to these pseudoproxy data over Europe. The resulting fields are then compared with the simulated fields (cf. Fig. 2).

The analog method uses observations, i.e., instrumental or proxy data, at selected locations during the target (reconstruction) interval. These data are then compared to a pool of analog data—here the full ECHO-G+MM5 data. The analogs best matching the proxy values best are then used as the reconstruction. The target interval was the period 1500–1990, corresponding to the period reconstructed by Pauling et al. (2006). The second method, BARCAST, uses a hierarchy of stochastic models for the climate field and the proxy response function. It uses Bayesian inference to estimate the joint distribution of the model parameters and the targeted climate field variables (Tingley and Huybers 2010, 2013). Its original formulation uses an AR(1) process for the climate field. While this is approximately true for temperature data, precipitation data had to be transformed to a normal distribution. Due to the high computational demands of the BARCAST algorithm, the gridded RCM data were reduced by about a factor 10 for each longitudinal and latitudinal direction. The reconstruction period was 1–1850, with a calibration interval of 1851–1998. The Spearman correlation between the RCM model output, i.e., the ECHO-G+MM5 simulation and the reconstructed field is also shown in Fig. 2 for winter precipitation.

3 Key Findings

The evolution of the simulated seasonal precipitation and temperature in two different and geographically non-overlapping regions over Europe is displayed in Fig. 1. We show also results for temperature because we investigate potential links between temperature and precipitation for our sample regions: the Alps (Fig. 1a, defined as the region between 5°–15°W and 44°–48°N) and Central Europe (Fig. 1b, between 3°W–15° W and 48°N–54°N). The figures display two important aspects: The first one relates to the GCM–RCM differences in the seasonal (2000-year) mean temperature (e.g., 0.44 °C for the Alps for winter between ECHO-G and ECHO-G+MM5) and precipitation (47 mm/month for the Alps for winter). This is indicative of a large model-structure uncertainty due to the specific physical configuration of the global and the RCMs including differences related to the treatment of convective precipitation and cloud-physics schemes. Most important is the better resolved topography in the RCM, leading for instance to lower temperatures and higher precipitation over the Alps (Fig. 1a).

A second aspect is the warming trend over the past two centuries. Winter temperatures in the late twentieth century show highest values within the last 2,000 years, while a period of above-normal summer temperatures compared to the mean of the entire 2,000 years is simulated around 1100. State-of-the-art reconstructions (PAGES 2k Consortium 2013) show also increased temperatures over Europe between 800 and 1,100 compared to the reference period 1190–1970. Precipitation variability at centennial time scales is high in winter, especially for Central Europe, where it can vary by 30 % (Fig. 1b). The relationship between temperature and precipitation in winter is weak, with cold/dry or warm/wet winters [correlations are $r = +0.26$ and $r = +0.39$ ($p < 0.01$)] for the Alps and Central Europe, respectively. In summer, the relationship is stronger [correlations are $r = -0.74$ and $r = -0.55$ ($p < 0.01$)] i.e., warmer/drier or colder/wetter summers.

Another intriguing aspect is connected to the large differences that occur between the output of the regional model and the global model. These differences occur in the long-term mean values and in the variability, indicating that the RCM generates substantial regional deviations from the large-scale driving conditions of the GCM. This is most noticeable for summer temperature and precipitation in Central Europe (i.e., precipitation differences around 1,200).

The differences between the two global-regional model combinations can be large (cf. Fig. 1a): the simulated winter precipitation trends in the Alps in recent centuries display opposite long-term trends, with progressing drier conditions in ECHAM5 +CCLM and wetter conditions in ECHO-G+MM5. In areas with less complex terrain (Central Europe), the difference between the regional models is less pronounced (cf. Fig. 1b). In these areas the usage of the direct GCM output—without further dynamical downscaling—appears a reasonable option.

Results of the PPE show in general high correlations in areas close to the proxy data for both methods. However, the complexity of precipitation complicates its (spatial) reconstruction, reflected by the lack of correlations in areas far away from proxy locations (i.e., the decrease of correlations with increase in distance from the proxy location in Fig. 2).

It is worth to be mentioned that this is an idealized experiment, where the pseudoproxies contain perfect information. Therefore these results should be viewed as the theoretically optimal result that can be achieved when applying the methods unchanged to real proxies. Despite its underlying simple assumptions the BHM using the BARCAST algorithm performs as well as the analog method.

Future work will concentrate on establishing potential relationships between the evolution of precipitation and changes in external forcings and internal variability. In this context, precipitation variability, also on longer time scales, is far more complex and spatially more heterogeneous and links with changes in external forcings are most likely to be weaker compared to temperatures. Moreover the methods used for the PPEs will be applied to real proxies related to the hydrological cycle including tree rings, varved lake sediments and speleothems. As indicated in the results correlations between temperature and precipitation in terms of warmer/ drier and colder/wetter summers could be identified that might have also influenced cultures and societies in pre-industrial Europe in the last 2,000 years (i.e., Tol and Wagner 2010).

References

Flückiger J, Monnin E, Stauffer B, Schwander J, Stocker T, Chappellaz J, Raynaud D, Barnola JM (2002) High resolution Holocene N_2O ice core record and its relationship with CH_4 and CO_2. Glob Biogeochem Cycles. doi:10.1029/2001GB001417

Franke J, Gonzalez-Rouco JF, Frank D, Graham N (2011) 200 years of European temperature variability: insights from and tests of the proxy surrogate reconstruction analog method. Clim Dyn 37(1–2):133–150. doi:10.1007/s00382-010-0802-6

Gagen M, Zorita E, McCarroll D, Young GHF, Grudd H, Jalkanen R, Loader NJ, Robertson I, Kirchhefer A (2011) Cloud response to summer temperatures in Fennoscandia over the last thousand years. Geophys Res Lett 38:L05701. doi:10.1029/2010GL046216

Gomez-Navarro JJ, Montavez JP, Wagner S, Zorita E (2013) A regional climate palaeosimulation for Europe in the period 1500–1990—part 1: model validation. Clim Past Discuss 9:1667–1682. doi:10.5194/cp-9-1667-2013

González-Rouco F, von Storch H, Zorita E (2003) Deep soil temperature as proxy for surface air-temperature in a coupled model simulation of the last thousand years. Geophys Res Lett 30:2116–2119

Jungclaus JH, Lorenz SJ, Timmreck C, Reick CH, Brovkin V, Six K, Segschneider J, Giorgetta MA, Crowley TJ, Pongratz J, Krivova NA, Vieira LE, Solanki SK, Klocke D, Botzet M, Esch M, Gayler V, Haak H, Raddatz TJ, Roeckner E, Schnur R, Widmann H, Claussen M, Stevens B, Marotzke J (2010) Climate and carbon-cycle variability over the last millennium. Clim Past Discuss 6:723–737. doi:10.5194/cp-6-723-2010

PAGES 2 k Consortium (2013) Continental-scale temperature variability during the last two millennia. Nat Geosci 6:339–346. doi:10.1038/NGEO1797

Pauling A, Luterbacher J, Casty C, Wanner H (2006) Five hundred years of gridded high-resolution precipitation reconstructions over Europe and the connection to large-scale circulation. Clim Dyn 26:387–405. doi:10.1007/s00382-005-0090-8

Rockel B, Geyer B (2008) The performance of the regional climate model CLM in different climate regions, based on the example of precipitation. Meteorol Z 17(4):487–498. doi:10.1127/0941-2948/2008/0297

Rutherford S, Mann ME, Osborn TJ, Bradley RS, Briffa KR, Hughes MK, Jones PD (2005) Proxy-based Northern hemisphere surface temperature reconstructions: sensitivity to methodology, predictor network, target season and target domain. J Clim 18:2308–2329

Seneviratne SI, Corti T, Davin EL, Hirschi M, Jaeger EB, Lehner I, Orlowsky B, Teuling AJ (2010) Investigating soil moisture-climate interactions in a changing climate: a review. Earth Sci Rev 99:125–161. doi:10.1016/j.earscirev.2010.02.004

Smerdon JE (2012) Climate models as a test bed for climate reconstruction methods: pseudoproxy experiments. Wiley Interdiscip Rev Clim Change 3:63–77. doi:10.1002/wcc.149

Tang Q, Leng G, Groisman PY (2012) European hot summers associated with a reduction of cloudiness. J Clim 25:3637–3644. doi:10.1175/JCLI-D-12-00040.1

Tingley MP, Huybers P (2010) A bayesian algorithm for reconstructing climate anomalies in space and time. part I: development and applications to paleoclimate reconstruction problems. J Clim 23(10):2759–2781. doi:10.1175/2009JCLI3015.1

Tingley MP, Huybers P (2013) Recent temperature extremes at high northern latitudes unprecedented in the past 600 years. Nature 496:201–205. doi:10.1038/nature11969

Tol RSJ, Wagner S (2010) Climate changes and violent conflict in Europe over the last millennium. Clim Change 99:65–79. doi:10.1007/s10584-009-9659-2

Werner J, Smerdon J, Luterbacher J (2013) A pseudoproxy evaluation of bayesian hierarchical modelling and canonical correlation analysis for climate field reconstructions over europe. J Clim. doi:10.1175/JCLI-D-12-00016.1

Zorita E, Gonzalez-Rouco JF, von Storch H, Montavez J, Valero F (2005) Natural and anthropogenic modes of surface temperature variations in the last thousand years. Geophys Res Lett 32(8):L08707. doi:10.1029/2004GL021563